AF306391

Glass Making in the Greco-Roman World

Studies in Archaeological Sciences 4

The series Studies in Archaeological Sciences presents state-of-the-art methodological, technical or material science contributions to Archaeological Sciences. The series aims to reconstruct the integrated story of human and material culture through time and testifies to the necessity of inter- and multidisciplinary research in cultural heritage studies.

Glass Making in the Greco-Roman World

Results of the ARCHGLASS Project

Edited by
Patrick Degryse

Leuven University Press

Published with support of

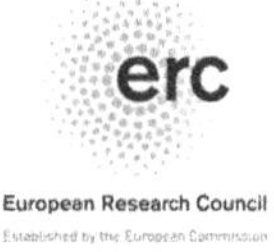

European Research Council
Established by the European Commission

ISBN 978 94 6270 007 9

D / 2014 / 1869 / 86
NUR: 682/933

Lay-out: Friedemann Vervoort
Cover: Jurgen Leemans

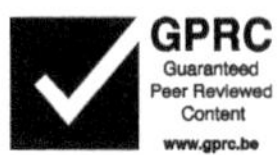

Preface

The ARCHGLASS "Archaeometry and Archaeology of Ancient Glass Production as a Source for Ancient Technology and Trade of Raw Materials" project, is a Seventh Framework Programme "Ideas" project funded under the European Research Council Starting Grant scheme. The main goals of the project were to develop innovative archaeometrical techniques to reconstruct ancient economies, and to apply these methods to gain new insights in the trade and processing of mineral raw materials used for glass making.

In effect, the project and this book presents a database of mineralogical and chemical compositions of possible raw materials (sand and flux) for primary glass making. From these data, the primary provenance of ancient natron glass can be derived. The focus of this project was on the Hellenistic-Roman world, investigating both production and consumer sites of glass from a chemical perspective, but integrating typo-chronological and archaeometrical information. Analyses of glass artefacts at consumer sites as well as a chemical characterisation of raw glass at primary production sites are presented.

Using an interdisciplinary approach, combining historical, archaeological and archaeo-metrical data, the occurrence of primary production centres of raw glass outside those known from archaeological excavation in Syro-Palestine and Egypt were investigated. In particular, the western Mediterranean area including the regions of Italy, Gaul, Spain (described by ancient authors as primary glass producers) and north-Africa were investigated. In this way, by developing new analysis techniques, an innovative archaeological interpretation of glass trade in the Hellenistic-Roman world can be given, and the possibility of integrating glass studies into the larger framework of Roman economy studies, now underrepresented, becomes possible.

The political situation in Northern Africa and the Middle East during the course of the project prohibited much fieldwork there. Only a small amount of samples from Israel, Egypt, Libya and Tunisia from previous field campaigns and from older, well-documented collections were studied. This makes the study of North Africa as a supplier of natron glass more difficult. Moreover, a database of analyses of possible raw materials and glass is never complete. The approach to the geological materials analysed here is based on regional survey, taking samples from different geological hinterlands with a wide spacing, but not going into very detailed sampling grids. Nevertheless, almost 300 samples of sand and flux and close to 400 analyses of glass are taken into account in this book. This analytical strategy makes the interpretation of provenance given in this book the best possible interpretation to the knowledge of the authors, at this point in time within the current state-of-the-art. The reconstruction of the primary origin of ancient natron glass presented here, hence refers to broad geological/geographical regions where glass was (likely) made, rather than to individual sites. The authors wish to make all this raw data available in open access format to all scholars in the field, to use in their own research, study or teaching… and judge for themselves. This book is written with archaeologists and historians in mind. The analysis procedures, for instance the newly developed methods of isotopic analysis, can be found in specialised journal papers. This book focuses on an archaeological interpretation of the exact data produced, especially in the conclusions at the end of each chapter, and the discussion in chapters six and seven.

Acknowledgements

Several museums, excavations and individual scholars provided samples and unravelled archaeological contexts for this study, for which we would like thank Lorenzo Appolonia, Rossella Arletti, Nizar Abu Jaber, Adam Aja, Bob Brill, Nick Cahill, Giacomo Chiari, Peter Cosyns, Wim Dijkman, Roald Doctor, Elizabetha Dotsika, Susanne Ebbinghaus, Danielle Foy, Patrizia Framarin, Ian Freestone, Caroline Jackson, Filomena Gallo, Domingo Gimeno-Torrez, Bernard Gratuze, Joe Greene, Hans Huismans, Susanna 'Suzy' Kirk, Roland Lamprichs, Veerle Lauwers, Marleen Martens, David Mattingly, Judith McKenzie, Luigia Melillo, Piero Mirti, Gianmario Molin, Patrick Monsieur, M. Pugès, Ziad al Saad, Mariarosa Salvatore, Valeria Sampaolo, Alberta Silvestri, Sofie Vanhoutte, Frank Vermeulen, Michael Vickers, Susan Walker, Erik Wetzels, the Sagalassos Archaeological Research Project, the Servei d'Arqueologia Urbana, Ajuntament (municipality) of Barcelona, the Getty Museum, the Getty Conservation Institute, the Harvard Art Museums, Harvard Semitic Museum. Femke Hasendonckx, Leander Geldhof, Maarten Diels were active as master students in this project. Elvira Vassilieva, Kris Latruwe, Steven Luypaers, Rita Giannini, Lieve Balcaen, Johan Honings and Kelly Domoney were essential for analytical support and interpretation of data.

Ian Freestone, Marie Dominique Nenna, Thilo Rehren, Frank Vanhaecke and Caroline Jackson are greatly acknowledged for feedback on ideas and discussions. Udo Haack, Jens Schneider and Philippe Muchez were essential in the design and development of this project, while Dennis Braekmans, Katherine Eremin, Andrew Shortland and Marc Walton have been friends, colleagues and *compagnons de route*, for which I cannot thank them enough.

Funding for this project was provided by several projects of the Fund for Scientific Research Flanders (FWO, G.002111.N, G-08-00291, G.A078.11), the KU Leuven (PFV/10/001) and the European Research Council under the Starting Grant scheme (grant agreement no. 240750). Annelore Blomme, Dieter Brems and Monica Ganio benefitted from a Research Assistant Grant of the Fund for Scientific Research—Flanders (FWO-Vlaanderen). Dieter Brems is currently working as postdoctoral fellow of the Fund for Scientific Research—Flanders (FWO-Vlaanderen). ERC allowed the training of a significant number of scholars, who will undoubtedly find their way into the international community in their research field.

Table of Contents

List of illustrations

List of Tables

Chapter 1

The archaeology and archaeometry of natron glass making

R.B. Scott, P. Degryse

Pliny the Elder, writing in the 1ˢᵗ century AD describes how glass was first made. *"There is a story that once a ship belonging to some traders in natural soda put in here and that they scattered along the shore to prepare a meal. Since, however, no stones suitable for supporting their cauldrons were forthcoming, they rested them on lumps of soda from their cargo. When these became heated and were completely mingled with the sand on the beach a strange translucent liquid flowed forth in streams; and this, it is said, was the origin of glass"* (Pliny *NH* 36.65; Eichholz, 1962: 151). Although the story is untrue, Pliny is describing Roman 'natron' glass. In fact, Pliny devotes the concluding chapters of his 36ᵗʰ book on Natural History to the discussion of silica and its practical use in the production of glass (Healy, 1999). He also describes several locations of suitable glassmaking sand *"That part of Syria which is known as Phoenicia and borders on Judea…This is supposed to be the source of the River Belus, which after traversing a distance of 5 miles flows into the sea near the colony of Ptolemais…The beach stretches for not more than half a mile, and yet for many centuries the production of glass depended on this area alone…Sidon was once famous for its glassworks, since, apart from other achievements, glass mirrors were invented there. This was the old method of producing glass. Now, however, in Italy too a white sand which forms in the River Volturno is found along 6 miles of the seashore between Cuma and Literno…Nowadays sand is similarly blended also in the Gallic and Spanish provinces"* (Pliny *NH* 36.65-66; Eichholz, 1962: 149-155). Pliny suggested that the sand from the River Belus, present-day Israel, was traditionally used in the manufacture of glass based on the 'old method' (Pliny *NH* 36.66). It is implied that the 'new method' of glass production required new materials and sand types, such as those from Italy (Freestone, 2008). Pliny also makes reference to the use of soda; in direct reference to glass, he merely says that it was added to the batch. He does however say that soda can be found in Media and in Thrace, near Philippi, but the latter is contaminated (Pliny *NH* 31.46; Jones, 1963: 443-445); he also says it can be found in Macedonia. He then explains how artificial soda is produced in

Egypt in large quantities and again, is of inferior quality (Pliny *NH* 31.46; Jones, 1963: 445). Pliny goes on to say that the soda-beds of Egypt are found in the regions around Naucratis and Memphis. Natural deposits of soda are well known at the Wadi el Natrun, c. 50km northwest of Cairo, Egypt and at al-Barnuj, Egypt (Freestone, 2008; Lucas, 1932; Shortland et al., 2006; Shortland, 2004). Since the former sources (Media, Thrace and Macedonia) are not mentioned in direct relation to glass manufacture (Freestone, 2008; Saguí, 2007), the use of these sources is still debated.

The Belus River is today known as the Na'aman Stream and it flows to the south of the city of 'Akko (Gorin-Rosen, 2000). Josephus, writing in the 1st century AD (Josephus *The Jewish War* 2.10.2), highlights the availability of glass making sand in the region around Ptolemais, but he does not specify where glass is actually being manufactured. He does mention both vitreous sand and vitreous matter; this could either be interpreted as glass-making sands or some form of raw glas or frit. Josephus does say that boats remove this material, implying that the final manufacture of the raw glass occurs elsewhere. He explains that this is a maritime area with a thriving port, which suggests that if glass is being made in the area, the mechanism for trade and transport is also present. Although there is evidence of glass production in the 'Akko region, the exact date of the sites are unknown, i.e. Roman or later (Gorin-Rosen, 2000). Strabo also mentions the location of glassmaking sands (Strabo *Geography* 16.2.25), again suggesting Syro-Palestine, and Ptolemais in particular, as the source, mentioning glass working as well as glass production. Like Pliny, he implies that different methods of producing glass are used in different areas and that these use different raw materials, but he does not specify whether these are 'old' or 'new' methods. Strabo also mentions the presence of soda lakes in Armenia (Strabo *Geography* 12.14.8), and nitre-beds near Momemphis and Naucratis in Egypt (Strabo *Geography* 17.1.23) but he makes no connection between these and the production of glass. Tacitus again mentions the Belus river as a source of glass making sand (Tacitus *Histories* 5.7). Several classical authors hence refer to the use of sand from the region of the Belus River. However, they all tend to suggest that the sand was collected in this region but the actual process of glass-making occurred elsewhere. Moreover, the Church and Brodribb (1864) translation of Tacitus adds to this interpretation because they specify that the beach *"furnishes an inexhaustible supply to the exporter"*.

Further literary evidence of glass production comes from the Edict on Maximum Prices issued by Diocletian in 301 AD. The text contains six lines referring to glass, and only Alexandrian and Judaean glasses are mentioned by name (Erim and Reynolds, 1973). It is unclear from the text whether the names refer to production locations for the glass or broader glass categories (as was suggested by Barag (Barag, 1987)). In relation to Judaean glass, the edict mentions 'vitri Judaici svirdis'. Svirdis has been interpreted as being either 'viridis' or 'subviridis' (Erim and Reynolds, 1973) meaning green or greenish glass. This does not mean, however, that all green glass or only green glass was made in Judaea, especially since the text also mentions Judaean plain glass cups and vessels. Likewise, it cannot be assumed from this that all Alexandrian glass was colourless, especially in light of the information supplied by Strabo where the Alexandrian glass workers speak of the vitreous earth needed to make coloured designs. Athenaeus, writing at the end of the 2nd century AD, says *"The inhabitants of Alexandria…work with glass, transforming it into cups of a wide variety of shapes and imitating the look of all the types of pottery imported from every corner of the world"* (Athenaeus *The Learned Banqueters* 11; Olsen, 2009: 261). The implication here is that the glass is worked but not manufactured from raw materials in Alexandria and the glass workers may also have tried to imitate pottery colours as well as design. Glass working is mentioned in various parts of the Historia Augusta (parts 18; 26; 29); mentioning glass working in Egypt and the taxes imposed on the production of glass (Magie, 1924; Magie, 1932). From these latter sources it can be concluded that glass was worked in Alexandria (and Judaea) but they shed little further light on the production location(s) of the raw glass.

The question of glass manufacture in the Roman Empire is further complicated by more recent antiquarian authors, who confuse the manufacture of raw glass with the secondary production of glass vessels, suggesting that glass was at one point or another produced in almost every area of the empire (Stevenson, 1914). From the historical texts it can be deduced that primary 'raw' glass was probably produced in Syro-Palestine, Egypt, Italy, Gaul, Spain and India, although the method of production may have varied in these locations.

1.1   The Archaeological Evidence

The archaeological evidence for the production of primary glass (i.e. fused from raw materials into a glass which is then broken into chucks and transported to workshops across the empire for shaping) in the Roman period is limited. Excavations have revealed that during the 4th – 8th centuries, large quantities of natron glass were made in a limited number of 'primary' centres in Egypt and Syro-Palestine (Brill, 1999, 1988; Nenna et al., 2000; Foy and Nenna, 2001; Foy et al., 2000; Freestone et al., 2002a, 2000; Picon and Vichy, 2003). For example, excavations at Bet Eli'ezer, Israel revealed the remains of 17 rectangular furnaces, dating to the 8th century AD (Freestone et al., 2008a, 2000; Gorin-Rosen, 2000).

While excavations at Apollonia, Israel revealed the presence of four furnaces, similar in style to those found at Bet Eli'ezer, dating to 6th-7th centuries AD (Freestone et al., 2008a; Gorin-Rosen, 1995, 2000; Perkins, 1951; Tal et al., 2004). More recently, evidence of primary and secondary production has been found at the site of Horbat Biz'a, c. 7km east of Bet Eli'ezer, although remains of the primary glass furnace itself have not yet been found (Gorin-Rosen, 2012). Glass furnaces were discovered on the shores of Lake Maryut, near Alexandria, Egypt, dating from the Imperial period to the 8th century AD (Nenna et al., 2000); while Roman glass furnaces of a 1st-2nd century AD date have been discovered in Egypt, at Wadi Natrun (Nenna, 2003, 2007; Nenna et al., 2000, 2005). However, these are of a different form to the later tank furnaces excavated in Palestine (Freestone, 2005; Nenna, 2007; Nenna et al., 2000). The tank furnaces of Israel were extremely large producing c. 8-9 tonnes of glass per firing (Freestone et al., 2000; Gorin-Rosen, 2000), but little further archaeological evidence of primary glass production in the 1st – 5th centuries has yet been found (Paynter, 2006). Although, it has been suggested that further earlier sites may exist in Syro-Palestine and Egypt (Foy et al., 2003; Nenna, 2003; Nenna et al., 2000, 1997). The furnace locations that have been found are located either near the favoured sand sources, such as the mouth of the Belus River, or close to the alkali sources, such as the Wadi Natrun, Egypt (Freestone et al., 2000; Nenna et al., 2000). This agrees with the historical sources, which say that glass was produced using the 'old' method in Syro-Palestine and Egypt. However, little evidence of primary glass factories have, as yet, been found in the other regions mentioned, for the earlier Roman period. Six glass factories from the late Imperial period (c. 4th century) have been suggested at Hambach, Germany (Wedepohl and Baumann, 2000; Wedepohl et al., 2011b), as well as a 2nd century AD tank furnace in a Roman military camp at Bonn (Wedepohl et al., 2011a).

Secondary glass workshops were more prolific across the empire. Also, historic records talk of glass workers being shipped from Syria and Judea to Rome (Fleming and Swann, 1999). Fleming and Swann do not provide a reference for this historic source, but archaeological evidence for the movement of glass workers does exist. This is in the form of names and sometimes origins of the workers recorded on the glass artefacts they produced (Price, 2005). For example, blown glass tablewares with the addition of 'the Cypriot' or 'of Sidon' (Nenna, 2007; Price, 2005) have been interpreted as meaning that the glass worker originally came from those areas. A further example given by Price (2005) is that of a 3rd century tombstone found in Lyons. It records the death of Julius Alexsander, master in the art of glass; he was born in Africa and was a citizen of Carthage (Foy and Sennequier, 1989; Price, 2005). Excavations at Pompeii revealed the presence of glass workshops which, based on the eruption of Vesuvius, must date to pre- 79 AD (De Francesco et al., 2010). Secondary production of glass dating from Imperial to early Byzantine times has also been found at Sagalassos, Turkey (Degryse et al., 2006a). The archaeological evidence for glassmaking in the western part of the Empire is scarce, but, by the late Roman period, glasshouses where vessels or window glass were fabricated were well established across this region (Foster and Jackson, 2010). For example, more than 70 workshops have been excavated in France (Foy et al., 2003; Nenna, 2007; Price, 2005), and more than 20 in Britain (Jackson et al., 2003a, b; Paynter, 2006; Price, 2005). The Roman glass workshops of Western Europe are far more common and often better documented than those of the Eastern Mediterranean (Stern, 2002), but even without the secondary workshops, there is strong evidence for glass production and trade. For example, Roman shipwrecks such as the Embiez or the Iulia Felix have been found carrying cargoes of glass and glass cullet (Silvestri, 2008). Shipwrecks along the coast of Israel have also been found carrying raw chunk glass, indicating trade by sea existed (Gorin-Rosen, 2000). Although cargoes of Roman glass are relatively uncommon, enough ship-wrecks have been found to confirm that a significant scale of long-distance, seaborne transport existed and that this was not confined to the more expensive, coloured or engraved glasses (Gibbins, 1991). It has been argued that most long-distance trade in the Roman world was related to the provisioning of the armies (Silvestri et al., 2006), but it has also been suggested that the formation of the Empire and the pacification of the Mediterranean basin in the time of Augustus created a new world market (West, 1932), which would have allowed an increase in all types of commercial activity. This coupled with the invention of glass-blowing in the 1st century made glass a widespread and commercially available product. Glass finds and references in historical documents reveal that glass was exported to many different areas of the Empire, often in large quantities (Thorley, 1969). *"The discovery that molten glass could be blown was nothing less than revolutionary. It was closely related to the equally momentous discovery that broken glass artefacts could be totally remelted, a breakthrough that kindled a literary response in the Flavian period (69-96) equal only to the excitement of Augustan poets about glassblowing"* (Stern, 1999: 450).

The archaeological evidence has led to the creation of two main models for glass production in the Roman Empire; local versus centralised. The early models of glass production were based on ideas about the structure of the ceramics industry (Freestone, 2005), i.e. it was hoped that the trade in glass vessels could be mapped based on the idea that a particular form or typology had a similar composition or came from a specific workshop. This idea anticipated that a workshop would produce glass of a constant composition which would distinguish it from the glass made in another workshop (Freestone, 2005). Each workshop would, in short, produce glass using the raw materials locally available. This 'local' model, similar to the Medieval model of glass production, would result in a large number of chemically distinct glasses being produced (Paynter, 2006). The second model, based on the archaeological evidence of surviving furnace sites (also of a later date than the Roman period), suggests that the glass was made in a small number of primary production centres, and the raw glass was then shipped to workshops all across the Empire for shaping (Freestone, 2005). This idea suggests that each primary factory could supply a large number of workshops, and that many different workshops could essentially produce a variety of different items from the same composition of glass. Likewise, a single workshop could receive glass from a number of primary factories (Freestone, 2005). This second model would therefore result in only a few chemically distinct groups (Paynter, 2006).

So, the two principal models for glass production and distribution in the Roman Empire are both based on models used for glass manufacture in later periods, i.e. Byzantine and Medieval. Although the local (Medieval) model does have supporters and several secondary workshops have been found, the evidence for the production of raw glass outside of the Near East in the early Roman period is still largely illusive. So, the more favoured model, based on the Byzantine system, suggests a small number of large primary factories were operating in the Near East and then transporting the raw glass across the Empire for working at secondary locations. *"In the absence of archaeological evidence, the extent to which the foregoing model is applicable to the ancient world beyond the Near East, where the factories have been discovered, can only be determined by the investigation of glass composition"* (Freestone et al., 2000: 67).

1.2   What is glass?

Glass is made from a combination of network formers, modifiers and stabilisers; silica (SiO_2) is the most common network former in ancient glasses. Unfortunately, pure silica glass requires temperatures in excess of 1700°C to produce (Matson, 1951; Morey and Bowen, 1925; Morey, 1938; Shelby, 2005), which was beyond the capabilities of ancient glass-making technology. In order to reduce the melting temperature of the silica a fluxing agent is needed, this can be either soda (Na_2O) or potash (K_2O). These network modifiers reduce the melting temperature (Matson, 1951; Morey and Bowen, 1925; Morey, 1938; Shelby, 2005). However, a glass made from only SiO_2 and Na_2O would be unstable and susceptible to damage from water, therefore, a stabiliser, such as lime (CaO), is also needed (Hodges, 1981).

Too little stabiliser and the glass has poor chemical durability, too much stabiliser and the glass is prone to devitrification. In this case a small amount of alumina (Al_2O_3) or magnesia (MgO) is beneficial as these can help prevent devitrification (Hares, 1984).

1.3   What is natron glass?

The glass manufactured during the Greco-Roman period was all soda-lime-silica in composition, the majority of these being made from a mineral soda (natron). This type of glass was made between the 5th century BC and the 9th century AD and is characterised by its low magnesium and low potassium values, also referred to as LMLK glass (Freestone, 2006; Henderson, 1985; Sayre and Smith, 1961; Wedepohl et al., 2011a). It can be differentiated from other types of soda-lime-silica glass which use plant ash as the main flux ingredient and are often referred to as HMHK glasses, due to their high magnesium and high potassium contents. A Roman 'natron' glass can therefore be recognised by concentrations of MgO and K_2O of less than 1.5%.

Suitable glassmaking sands are hard to find, they need to be high in silica and relatively free from impurities. They should also be relatively calcareous (high in lime) or extra lime must be added to the glass batch (Silvestri et al., 2006). Certain beach sand, containing lime in the form of crushed shells, was particularly suited to the purposes of making natron glass (Aerts et al., 2003, 2000; Brems et al., 2012a; Brill, 1988; Sode and Kock, 2001). Colourless glass became popular in the Roman period and this would have made low-iron, high purity sand essential for the glass industry (Jackson, 2005) in particular, for the production of colourless glasses. Most of the suitable glassmaking sands are thought to be found in the Eastern Mediterranean region. These typically contain around 9% CaO, 3-5% Al_2O_3 and less than 1% MgO (Aerts et al., 2000; Brill, 1988).

The soda used in Roman glass was a mineral form known as natron. The term natron is usually used to describe polyphase evaporitic lake deposits that are rich in sodium carbonates (Shortland et al., 2006). Most of the carbonate is in the form of the mineral trona; however, the natron deposits are hardly ever pure carbonates and usually also contain significant amounts of chlorides and/or sulphates. These are highly undesirable because of their limited reaction with silica at the temperatures achieved in the traditional glassmaking furnaces (Freestone, 2006; Gorin-Rosen, 2000). Virtually all Roman glass contains 0.5-1.2% Cl and 0.2-0.5% SO_4 (Bateson and Turner, 1939; Brems, 2012; Gerth et al., 1998; Shugar and Rehren, 2002). During the glassmaking process, excess sodium chloride and sulphate forms a second melt phase on the surface of the molten glass. This immiscible 'galle' could be skimmed off or discarded after the glass has cooled (Freestone, 2006; Gerth et al., 1998; Shugar and Rehren, 2002; Tanimoto and Rehren, 2008).

Since glass needs some form of stabiliser, Roman glasses usually contain between 5 and 10% lime (CaO). The major source of lime in these glasses would have been calcium carbonate and this was either added deliberately to the glass batch, as an independent component, or accidentally included, as particles of shell or limestone in the sand (Freestone, 2006).

1.4   Glass Provenancing

In order to clarify questions over the structure of the glass industry in the Roman period, it is essential to provenance the origin of the glass and its raw materials. This information can then be used in conjunction with the archaeological glass assemblages to develop and interpret patterns of trade and use within the Roman period. In its wider context, this information can also help to shed light on the cultural interactions necessary in order for the economic impact of the Roman glass industry to be assessed. The idea that an artefact can be matched to its geological source location is not new. It has often been used to form the basis of many archaeological theories relating to the migrations of peoples, as well as social interactions and exchanges (Pillay, 2001). The determination of the provenance of an object relies on the assumption that there is a measurable scientific property which will link an artefact to a particular source or production location. In the case of glass, provenance is used to refer to the origin of the raw materials and/or the place where the raw glass material was made (Degryse et al., 2010). This is in contrast to the art world use of the term 'provenance' which refers to the history of the artefact (Wilson and Pollard, 2001). In theory, the materials used in the manufacture of the glass are matched to the correct geographical source location. This relies on being able to identify the correct signature of the raw material, and that this signature is inherent from the geological source from which it originated. It is also assumed that the geological signature is not transformed physically or chemically during the manufacturing process (Ixer, 2007). Likewise, it is assumed that each raw material has an individual geological signature, for example, sand sourced from Italy will have a different signature to sand sourced from Israel. The glass manufacturing and working processes can also leave a signature in the finished product, for example, glass can be contaminated by the crucible material (Jackson et al., 2003b). Since the chemical properties of an object are seen as characteristic of the raw material source, the composition of the artefact is effectively a 'chemical fingerprint' (Wilson and Pollard, 2001). In order to provenance an item there are a number of assumptions which are made:

- The chemical signature/fingerprint of the raw materials varies between geological sources and that those signatures can be related to geographical locations. The inter-source variation needs to be greater than the intra-source variation for the source to be successfully identified.
- Some of the chemical characteristics of the raw materials are present in the final product either unchanged or changed in a predictable way.
- That any mixing of raw materials which occurs during the production of an artefact can be reasonably accounted for.
- That any post-depositional processes, i.e. weathering, that may affect the chemical signature can be accounted for, or suitable allowance can be made (Wilson and Pollard, 2001).

In other words, the composition of the different glass samples are analysed and compared, and any differences between the samples are determined, i.e. are distinct compositional groups forming? Once the group(s) have been established, the data is compared to existing data groups of known provenance. The results can then be compared to data for specific raw materials from specific sources. The latter part, however, presupposes a knowledge of all the characteristics of all the possible sources of a particular raw material, which is rarely the case (Harding et al., 2004). Homogeneity between sources can make provenancing difficult, therefore, the technique will only ever confirm that an item did <u>not</u> come from a location.

The provenancing of glass is further complicated by the fact that glass is a complex material and the relationship between the raw materials and the finished product is not a simple one. During glass melting many characteristics of the raw materials, such as mineralogy, grain size and shape, are lost so that only bulk chemical data is available (Brems, 2012). Roman vessel glass from the 1st – 6th centuries AD shows a very similar chemical composition irrespective of the temporal or geographical origin of the material (Aerts et al., 2000). This is particularly the case for the major element composition, for example, almost all Roman natron glass has a CaO content of c. 7.48% (Foster and Jackson, 2009). However, it should be noted that different authors sometimes quote a different mean average composition, for example, Wedepohl and Baumann (Wedepohl and Baumann, 2000) use 6.41 ± 0.9% CaO. For the rest of this volume, the 'typical' Roman composition has been taken from Foster and Jackson (2009). So, analysis of assemblages consisting of vessels of different forms, believed to have been produced in different locations, are generally indistinguishable compositionally

(Baxter et al., 2005). Likewise, basic utilitarian vessel forms have rarely been attributed with a specific origin and so the exact provenance of these is often difficult to determine (Gibbins, 1991). Supporters of both production models have attempted to explain this phenomenon. Those in favour of the local production model claim that the compositional consistency is due to the use of similar, strictly controlled recipes and production techniques or the reuse of glass in the form of cullet (Baxter et al., 2005). While the supporters of the centralised production model say that the compositional homogeneity is due to the use of raw materials from a limited number of locations.

The provenancing of the chemical fingerprint of Roman natron glass relies on the assumption that the glass was produced in discrete centres using standardised raw materials and manufacturing techniques. It also relies on the assumption that the signature of each glass making centre is individual (Jackson, 2012). As mentioned previously, the major elemental signature of Roman natron glass is fairly homogeneous. If the raw glass was manufactured from a relatively pure silica source with only minor/trace level impurities and natron (also low in impurities), then this is likely to cause a homogeneous signature (Jackson, 2012). Likewise, the homogeneous signature of the Roman glasses could be promoted due to the practice of recycling (Degryse and Shortland, 2009; Degryse et al., 2006a). Therefore, something more than just the bulk chemical data of the glasses is needed to enable successful provenancing. Several attempts have been made to use a variety of techniques to provenance glass, these include trace element analysis, rare earth element patterns and isotopic signatures (Degryse and Schneider, 2008; Degryse and Shortland, 2009; Freestone et al., 2003; Henderson et al., 2005; Shortland et al., 2007; Wedepohl and Baumann, 2000). In particular, Sr-Nd isotopic signatures have been used to indicate that the majority of the Roman glass found in Europe came from the Eastern Mediterranean region (Freestone et al., n.d.). Isotopic and trace element analyses are promising as tracers for the raw materials used in glass because they reflect the variations around the Mediterranean Sea which result from the differing geological environments (Brems, 2012). The chemical and isotopic analysis of the glass can, therefore, provide information over the origin of the raw materials used to produce the glass. This in conjunction with a comparison of the compositional data between archaeological sites can potentially reveal patterns in the production and trade of glass (Schibille, 2011). So, in order to fully understand glass manufacture in the Greco-Roman world, in the absence of identifiable primary glass production locations, a structured analysis of glass must be undertaken using contextual, chronological, typological and technological evidence from a wide group of assemblages (Baxter et al., 2005).

This fundamentally leads to two methods of provenance determination; the first compares the composition of the unknown samples with the composition of known material in order to narrow down the potential source origins. The second method uses the fundamental geological properties of the artefact to predict a potential origin in the absence of comparative material (Freestone et al., n.d.). The former method relies on the existing chemical and isotopic data collected from various Roman glass assemblages, such as the composition of glass collected from the Roman furnaces in Egypt (Nenna et al., 2000); the small scale production from York, England (Bingham and Jackson, 2008; Jackson et al., 2003b); or the detailed evidence from the 4th century glass possibly made at Hambach, Germany (Wedepohl and Baumann, 2000; Wedepohl et al., 2011a); as well as the detailed studies undertaken on 4th-5th century HIMT glass and 4th-8th century Levantine glass (Freestone, 2001; Freestone et al., 2002b). The second method utilises isotopic studies, particularly of Sr and Nd, to infer the likely locations of glassmaking sands and therefore the location of the primary furnaces in which the raw glass was made (Freestone et al., n.d.). It should be remembered, that *"to understand how glass can be related back to the furnaces in which it was made, the origins of all the components in the glass must be understood"* (Freestone, 2005: 008.1.3).

1.5   The ARCHGLASS project

The ARCHGLASS "Archaeometry and Archaeology of Ancient Glass Production as a Source for Ancient Technology and Trade of Raw Materials" project aimed at developing innovative archaeometrical techniques for provenancing, and to apply these methods to gain new insights in the trade and processing of mineral raw materials used for glass making. From our archaeological and archaeometrical analyses, it is clear that suitable sands for natron glass making are rare. Only a few regions and geological situations so far have been recognized as having suitable silica sands for making natron glass with a Greco-Roman composition, as discussed in Chapter 2. From the study of sand deposits around the Mediterranean, and from major and trace element and Sr-Nd isotope ratio analysis of ancient primary glass, the most important chemical provenance indicators of a primary glass origin can be derived, as discussed in Chapters 3 and 4. In terms of major to trace element composition, the Al_2O_3, TiO_2 and Zr contents of ancient glass are the most relevant indicators for a differing primary origin of the silica raw material used. In terms of isotopic analysis, the Nd isotopic composition of ancient glass is able to distinguish between different silica sources used. B isotopic analysis was proven to be a good indicator to trace the origin of the natron flux used in glass making, as discussed in Chapter 5. Combined, these chemical indicators have the most potential to source primary glass making in the Greco-Roman world to different geological-geographical areas around the Mediterranean basin, as discussed in Chapter 6. In conclusion, several phases in the location of glass making in the Hellenistic-Roman world are discussed in Chapter 7.

Chapter 2

Western Mediterranean sands for ancient glass making

D. Brems, P. Degryse

Ancient natron glass was essentially a mixture of three components: quartz sand, fluxed with a soda-rich mineral matter called natron and stabilised with lime. The natron used was a fairly pure soda source, being relatively free of potash and magnesia (Brill, 1988; Brill, 1999; Shortland, 2004). The major source of lime would have been calcium carbonate, either added deliberately to the glass batch as a separate component or accidentally as particles of shell or limestone in the sand (Freestone, 2006). Silica sand was by far the major component of ancient glass. Glass factories were probably built in the vicinity of suitable sand sources since the transport of vast quantities of sand would have been costly. Therefore, when trying to trace the origin of ancient glass, the sand source is an excellent place to start. If sand raw materials would have been transported to glass production sites in other areas, the geochemical characteristics of the glass would still point to the origin of these raw materials and not to the source of the raw glass as such. To our knowledge, no evidence for the transportation of sand raw materials for glass production has ever been found. But then again, the presence of sand in for example a shipwreck would not strike the discoverer as abnormal and would probably be overlooked.

Pliny the Elder's Natural History has been cited numerous times in studies of ancient natron glass production. Pliny describes the production of glass using sand from the beach near the mouth of the River Belus (Israel) and the coastal strip between Cuma and Liternum near the River Volturno (Italy) (see also chapter 1). The suitability of the River Belus sand for glass production has been proven by several studies (Turner, 1956; Brill, 1988; Vallotto and Verità, 2002). The Volturno sands, however, contain significant amounts of feldspar and augite, which results in relatively high aluminium and iron contents, making the sand unsuitable for glass production in its original state (Turner, 1956; Brill, 1999; Vallotto and Verità, 2002; Silvesti et al., 2006). Pliny further mentions that '*Nowadays sand is similarly blended also in the Gallic and Spanish provinces*' (Pliny *NH* 36.66; Eicholz, 1962: 155), suggesting that glass was made from raw materials in France and Spain.

However, no excavations have found evidence to support this. Furthermore, the suitability of these sands has never been evaluated. Previous archaeometrical work has not contradicted a possible western Roman production (Degryse and Schneider, 2008; Brems et al., 2012b). Pliny also noted that fossil, i.e. inland, sands could be added to the glass batch. Indeed, a mixture of mature, quartz-rich quarry sand and deliberately added shell material or limestone might be suitable to produce glass.

In this chapter, the possible existence of a Roman primary glass industry in the western part of the Mediterranean is investigated on the basis of the occurrence of suitable sand raw materials. Beach sands from the western Mediterranean Sea (focussing on Spain, France and Italy) are evaluated for their suitability for glass production by calculating the composition of hypothetical glasses made from these sands and comparing them to the composition of archaeological Roman natron glass. We focus on the suitability of readily available beach sands as the source of silica in ancient glass making.

2.1   The sand survey and analysis

Backshore sediment samples were collected from 178 sandy beaches along the coasts of Spain, France and Italy. Their geographic coordinates are given in Appendix A. The sampling locations were selected to represent sediment derived from all major geological units based on the geological maps of Spain (Instituto Technológico Geominero de España, 1994; Institut Geològic de Catalunya, 2006, 2011), France (BRGM, 2003) and Italy (Agenzia per la Protezione dell'Ambiente e per i Servizi Tecnici, 2004). The variability of the sands' mineralogical and chemical composition on a single beach was extremely limited, and within analytical error of the methods used (Pauwels and Brems, 2014).

The major and minor elemental compositions of the sands were obtained by ICP–OES chemical analysis. Complete analytical procedures are reported in Brems et al. (2012b, c).

In order to evaluate their usability as a raw material for glass production, the compositions of the hypothetical glasses made from these sands were calculated. This was done by removing the weight fraction lost on ignition (LOI), raising the soda concentration to 16.63% (the average Na_2O content of Roman glass; Foster and Jackson, 2009), thus simulating the addition of pure trona, and renormalising the remaining elements to 83.37% (Fig. 2.1). The calculated compositions of the hypothetical glasses were compared to the composition of Roman natron glass. Several different glass groups have been identified, such as Levantine 1, Levantine 2,

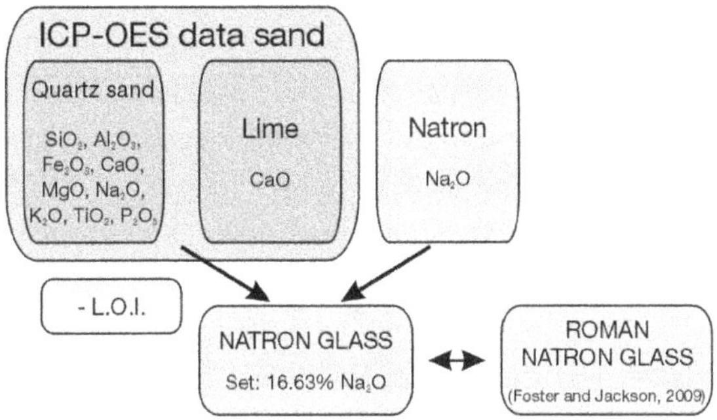

Figure 2.1:
Schematic representation of the method used to calculate the composition of glass that can be produced from the analysed beach sands after the addition of pure Na_2O to the glass batch.

Egypt 1, Egypt 2 and HIMT (Nenna et al., 1997; Freestone et al., 2000, 2002b, 2005; Foy et al., 2003; Freestone, 2006). For comparison, we used the average values given by Foster and Jackson (2009) for first to fourth century natron glass from the eastern and western part of the Roman Empire. Compositional ranges were chosen as the average values ± two times the standard deviation, thus comprising 95% of the dataset. This resulted in the tolerance ranges shown in Table 2.1. It must be noted that the assumption of a normal distribution for all chemical elements is not really correct. For elements such as Fe_2O_3, MnO and TiO_2, the use of the average concentrations and standard deviations to define the compositional ranges leads to

	Mean	StDev	Lower limit	Upper limit
SiO_2	69,54	2,53	64,48	74,60
Na_2O	16,63	1,50	13,63	19,63
K_2O	0,75	0,24	0,27	1,23
CaO	7,48	1,18	5,12	9,84
MgO	0,59	0,29	0,01	1,17
Fe_2O_3	0,62	0,48	0,00	1,58
Al_2O_3	2,59	0,38	1,83	3,35
P_2O_5	0,12	0,05	0,02	0,22
MnO	0,73	0,74	0,00	2,21
TiO_2	0,13	0,14	0,00	0,41

Table 2.1:
Tolerance ranges for the chemical elements. The ranges are chosen as the average ± two times the standard deviation using the values of Foster and Jackson (2009). All values are in wt%.

unrealistic (i.e., negative) lower limits. For these elements, the lower limits were adjusted to zero. Although all Roman glass contains at least trace concentrations of these elements, for our purposes these tolerances ranges are sufficiently accurate. For Late Roman and Byzantine glass, the ranges for CaO (4 – 11%) and Al_2O_3 (1 – 4.5%) would be somewhat wider (Freestone, 2006). Specific glass compositions like the HIMT (High Iron Manganese and Titanium) type were not considered.

For sands containing too little lime to produce an acceptable Roman glass a second calculation was carried out. Here, the CaO content was raised to the average CaO content of Roman glass (7.48%; Foster and Jackson, 2009) to model the deliberate addition of extra shell or limestone to the glass batch (Fig. 2.2).

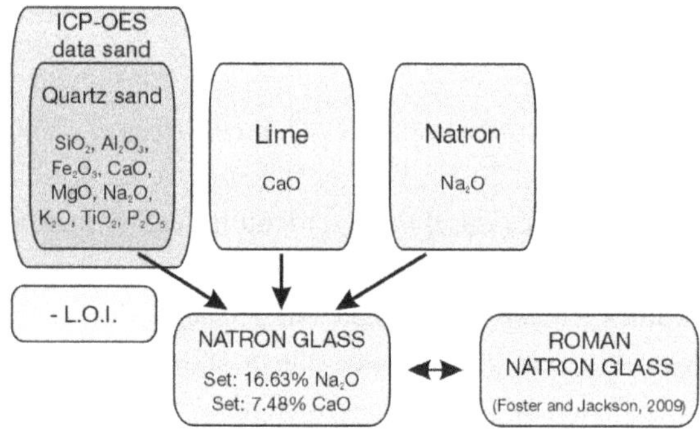

Figure 2.2:
Schematic representation of the method used to calculate the composition of glass that can be produced from the analysed beach sands after the addition of Na_2O and extra CaO to the glass batch.

The major and minor element concentrations and the LOI results of the analysed beach sands are listed in Appendix A. The compositions of theoretical glasses recalculated after the addition of 16.63% Na_2O are listed in Appendix B and shown in Fig. 2.3. The results are shown from west to east along the Mediterranean coast, from the Spanish–Portuguese border to the border between Italy and Slovenia. An extensive and detailed description of the results and the relation between the composition of the beach sands and the local geology is published elsewhere (Brems et al., 2012b, c).

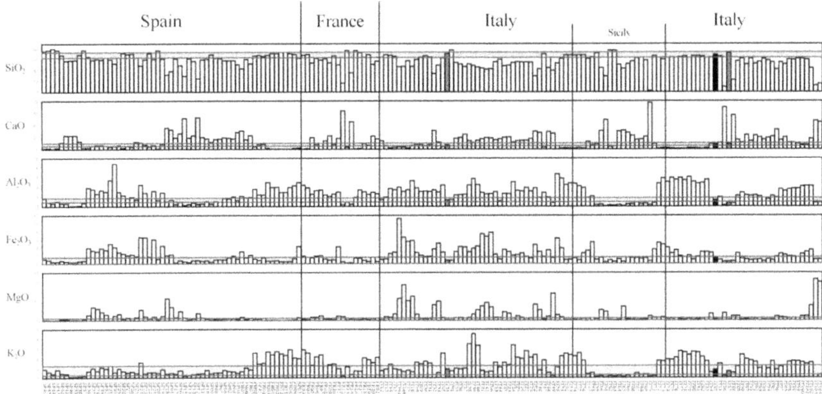

Figure 2.3:
Histograms showing the elemental composition of hypothetical glasses that could be made from the beach sand samples analysed after raising the Na_2O content to 16.63%. The light grey boxes represent the compositional ranges of Imperial Roman natron glass (Foster and Jackson, 2009). The sand sample corresponding to Roman natron glass for all major and minor elements (IT85) is shown in black. Sand samples IT34 and IT87, which differ from typical Roman glass in one element, are indicated in dark grey. The results are shown from west to east along the coast. All values are in wt%.

2.2   Suitable sands in the western Mediterranean

Present-day beach sands are evaluated for their suitability for Roman glass production. Any archaeological conclusions drawn from the results obtained, rely on the assumption that the composition of beach sands along the investigated coasts has not changed during the last two millennia. Under the Mediterranean climatic conditions which prevail in the study area, chemical weathering is practically absent, causing little or no change to the composition of detritus during weathering and erosion (Johnsson, 1993; Garzanti et al., 2002). Furthermore, most of the rivers delivering sediment to the studied beaches have moderate to high reliefs and short courses (Macklin et al., 1995; Poulos and Collins, 2002; Liquete et al., 2005). Detrital matter undergoes very limited compositional change during transport in such river systems and as a result, the composition of the beach deposits is generally a very good reflection of the dominant rock types in the hinterland (Cavazza et al., 1993; Critelli et al., 1997, 2003, Garzanti et al., 2002). Critelli et al. (2003) investigated the composition of sand deposits from fluvial and beach environments along the Atlantic and Mediterranean sides of the Betic Cordillera in southern Spain. They were able to define three distinctive sand petrofacies corresponding to the major geological source areas.

Similar investigations in the present study area were carried out along the coastal stretches of the French Mediterranean (Duplaix, 1972 and references therein), Liguria (Garzanti et al., 1998, 2004), Elba (Picard and McBride, 2007), Tuscany–Lazio–Campania (Garzanti et al., 2002), Calabria (Le Pera and Critelli, 1997) and Romagna–Marche (Rizzini, 1974). From these studies and the results discussed in Brems et al. (2012b, c) it can be deduced that sand compositions reflect tectono-morphologic littoral provinces and that the composition of beach and fluvial sands is very strongly controlled by the source land and the local geology. Since no major changes in the dominant source rocks are likely to have occurred over this short (geological) time scale, it seems reasonable to assume that the current sediments are still representative for beach sands in Roman times. Bellotti et al. (2004) previously noted that only very subtle compositional and mineralogical changes could be detected in sediments deposited in small delta systems through the late Holocene.

Other arguments supporting the stability of beach sand compostions are that, apart from the deltaic regions of the major rivers such as the Ebro, the Rhône and the Po, the fluvial networks have not significantly changed over the past 2000 years (see also Rizzini, 1974; Gandolfi et al., 1982; Viñals and Fumanales, 1995; Guillén and Palanques, 1997; Santisteban and Schulte, 2007; Larue, 2008; Rey et al., 2009). More information on the structural buildup and stratigraphy of the Ebro (Maldonado, 1972, 1975; Aloisi and Duboul-Razavet, 1974), Rhône (Oomkens, 1970; Aloisi and Duboul-Razavet, 1974; Vella et al., 2005) and Po (Nelson, 1970; Amorosi and Milli, 2001) delta systems can be found elsewhere. Next to the river systems, sea levels in the Mediterranean area have also been relatively stable over this time period with maximum variation estimated to be less than 1 m over the last 3000 years (Flemming, 1969; Lambeck and Bard, 2000; Antonioli et al., 2002; Lambeck et al., 2004; Sivan et al., 2004). In the survey, care was taken for any dredging or modern beach construction. However, in the Mediterranean, such process will involve only a limited transport of sand material, and maximal use of local deposits, as long distance transport is very costly.

2.2.1   Suitable glassmaking sands

From the results shown in Fig. 2.3 it can be concluded that sands suitable for the production of Roman natron glass are not common. After fluxing them with pure trona, only 1 out of the 178 analysed beach sands (IT85) would produce a glass with elemental compositions within the ranges of Imperial Roman natron glass for all major and minor elements. Two samples (IT34, IT87) differ from Roman glass in one characteristic. The rest of the sands analysed are unsuitable for Roman natron glass production in their present form, the determinant factors often being the insufficient SiO_2 content, the high Al_2O_3 and Fe_2O_3 levels and either too low or too high CaO contents (Fig. 2.4a). It must be noted that some of these sands can certainly be melted into a glassy material, but the composition of that glass would differ significantly from the typical Roman natron glass composition.

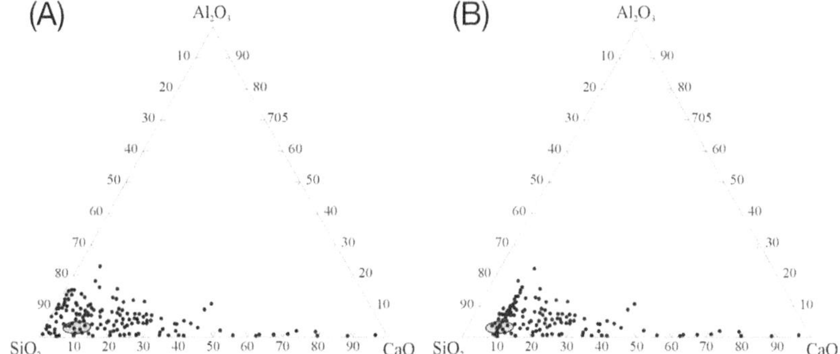

Figure 2.4:
SiO_2-Al_2O_3-CaO ternary diagram showing the compositions of the calculated glasses that could be produced from the analysed beach sands. The data are normalised to 100% to fit the diagram. The shaded area represents the compositional ranges of Imperial Roman natron glass (Foster and Jackson, 2009). (a) Hypothetical glasses made with the addition of pure natron. (b) Hypothetical glasses made with the addition of pure natron and additional calcium carbonate. Reprinted from Journal of Archaeological Science, Vol 39, Brems, D., Degryse, P., Hasendoncks, F., Gimeno, D., Silvestri, A., Vassilieva, E., Luypaers, S., Honings, J., Western Mediterranean sand deposits as a raw material for Roman glass production, 2897-2907, Copyright (2012), with permission from Elsevier.

The best Roman-type glassmaking sand occurs in the Basilicata region in SE Italy, between the mouth of the River Basento and the River Bradano (sample IT85). After fluxing them with pure natron, these sands would produce a glass approaching typical Roman natron glass: 69.6% SiO_2, 16.6% Na_2O, 8.5% CaO, 2.4% Al_2O_3, 1.3% Fe_2O_3, 0.5% MgO, 0.8% K_2O, 0.08% MnO, 0.10% TiO_2 and 0.04% P_2O_5. In the area southeast of Brindisi, in the Puglia region in SE Italy (sample IT87), locally made glass would be unusually low in Al_2O_3. But the rest of the chemical elements would fall within Roman ranges: 72.6% SiO_2, 16.6% Na_2O, 8.1% CaO, 1.1% Al_2O_3, 0.5% Fe_2O_3, 0.3% MgO, 0.6% K_2O, 0.02% MnO, 0.04% TiO_2 and 0.03% P_2O_5.

Both IT85 and IT87 are composed of the eroded material from Pliocene–Pleistocene sedimentary rocks. These sequences of sandstones, pelites, conglomerates and limestones further crop out in large areas in Basilicata and several, more scattered, parts of southeastern Apulia. Other beaches along these coasts may contain sands which are also suitable to make Roman glass depending on the ratios of quartz, calcite and feldspar derived from the local hinterlands. Diels et al. (2014) carried out a more systematic sampling of beach sands in these areas to map the distribution of suitable glassmaking sands in detail. Their results indicate that suitable sands for the production of Roman natron glass are indeed restricted to the northern part of the Gulf of Taranto, between Metaponto and Palagiano, and the strip of coast between Brindisi and Torre Rinalda. Other sand deposits in the area contain too much calcite resulting in CaO concentrations higher than those typical of Roman natron glass.

Also in the Toscana region (Italy), between Piombino and Follonica in the western part of the Follonica Gulf (sample IT34), a good glassmaking sand was found. Except for the low P_2O_5 concentration, all analysed major and minor elements fall within the range of Roman glass: 71.4% SiO_2, 16.6% Na_2O, 7.2% CaO, 2.6% Al_2O_3, 0.9% Fe_2O_3, 0.3% MgO, 0.9% K_2O, 0.07% MnO, 0.07% TiO_2 and 0.017% P_2O_5.

These three sands which would produce glasses of acceptable composition are all mainly derived from the recycling of older sedimentary successions. This reflects the importance of polycyclic chemical and mechanical weathering in the maturation of sediments. The necessary amount of CaO is brought to the sand either through calcareous fragments from minor (but important) limestones or marls in the local hinterland (IT85 and IT87), or through contributions of shell fragments naturally included in the sand (IT34). Detritus coming from the weathering and erosion of primary magmatic or metamorphic rocks are generally very rich in feldspar and other aluminosilicates, resulting in very high Al_2O_3 concentrations. This is for example the case for beach sands derived from the granitic plutons along the Catalonian Coastal Ranges and the metamorphic and magmatic basement cores in the Calabria Region. Decomposition of carbonate rocks, which are abundantly present throughout the study area, produces sands very rich in calcite and/or dolomite. For the sediments along the north Italian beaches this is most distinct. Here, abundant limestone and dolostone lithic fragments, derived from the Dolomite mountains, give very high CaO and MgO values.

2.2.2   Suitable lime-deficient sands

54 of the sands analysed contain too little CaO to produce a typical Roman glass. If this is the case, pieces of shell or limestone can be added to the glass batch, as was already suggested by Pliny the Elder in his Natural History. In a second calculation the CaO contents of the hypothetical glasses containing too little CaO were therefore raised to 7.48%, the average CaO content of Roman glass (Foster and Jackson, 2009), to model this deliberate addition of extra lime. The resulting calculated glass compositions are given in Appendix C and Fig. 2.5. The addition

of extra lime to the glass batch can be very beneficial for the resulting glass composition (Fig. 2.4b). Although the quality of most examined sands is improved in this way, only three samples (SP46, SP20 and FR16) could be brought to within compositional ranges of Roman glass for all major and minor elements. Another five of the sands analysed (SP45, SP43, SP22, FR17 and IT01) could be improved to match all but one elemental characteristic of Roman natron glass.

Sands from the western (IT34) and eastern (IT01) side of the Follonica Gulf, between Piombino and Punta Ala, are very similar in composition except for the concentration of CaO. IT34 is much richer in calcite grains and shell fragments, which deliver enough CaO (7.2%) to produce a stable glass. Sample IT01, on the other hand, contains only traces of calcite and shell and is therefore too low in CaO (0.6%). Furthermore, the SiO_2 level in this sample was too high (78.1%). After the addition of extra calcium carbonate to sands from the eastern part of the Follonica Gulf (IT01) they would produce very similar glass as sand from the western part of this region (IT34). Just like with glasses made from IT34, all elements would fall within the range of Roman glass except for the low P_2O_5 (0.009%).

Beach sands very rich in quartz occur along the Maures-Tanneron Massif in the Provence (FR16, FR17). Next to the abundant quartz grains, the sediments here also contain important amounts of feldspar and micaceous metamorphic rock fragments. This mineral composition makes these sands initially sufficient or slightly too high in SiO_2 and too high in Al_2O_3. Because of the lack of calcite, there is a deficiency in CaO (0.4 – 0.9%). After the addition of extra lime to the glass batch, the SiO_2 and Al_2O_3 concentrations have dropped to tolerable levels for sample FR16. This sand from the Bay of Hyères would now produce a glass of which all analysed elements are comparable to Roman glass. Sand from the Bay of Cavalaire, a bit further to the east (FR17), had sufficient SiO_2 in its original state, but due to the higher amounts of feldspar and staurolite it contained excessive Al_2O_3 (6.6%). Because of this very high Al_2O_3 concentration, the extra dilution effect from the added CaO was not enough to lower the Al_2O_3 content to acceptable levels. Apart from the Al_2O_3 all elements fall within tolerable ranges.

Along the southern coast of Murcia and the eastern coast of Almeria (SE Spain), the metamorphic Alpujárride Complex provides abundant metamorphic lithic fragments to the beaches. However, at a small beach near Águilas (SP20) and near the Antas River mouth (SP22) sediments are mainly derived from local Cenozoic siliciclastic sedimentary rocks overlying the Alpujárride Complex. These sands are relatively rich in quartz grains and contain only minor feldspar and lithic fragments with aluminosilicates. The beach sand near Águilas (SP20) contains only traces of naturally occurring calcite or shell, but after the addition of an extra source of lime to the glass batch, this sand would produce very good

Roman natron glass. Sediments provided by the Rio Antas (SP22) contain a few more impurities. The CaO (4.3%) content of the initially calculated glass was higher than that of glass made with sand from Águilas, but still too low. After the addition of a little extra lime the produced glass would be a little too high in Fe_2O_3 (1.6%), but all other elements would occur in acceptable concentrations.

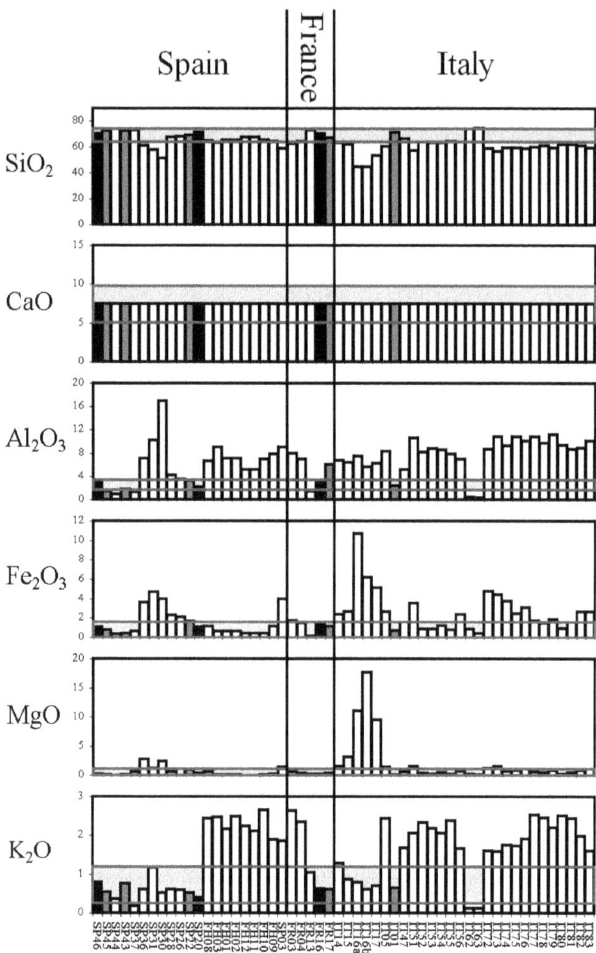

Figure 2.5:
Histograms showing the elemental composition of hypothetical glasses that could be made from the sand samples containing insufficient lime in their original state after raising the Na$_2$O content to 16.63% and the CaO content to 7.48%. The light grey boxes represent the compositional ranges of Imperial Roman natron glass (Foster and Jackson, 2009). The sands that could be used to produce typical Roman natron glass (SP46, SP20 and FR16) are shown in black. Sand samples producing glass which would differ from Roman glass in one element (SP45, SP43, SP22, FR17 and IT01), are indicated in dark grey. The results are shown from west to east along the coast. All values are in wt%.

Long-lasting mechanical weathering of sediment grains derived from igneous and metamorphic rocks from the Iberian Massif in the Guadiana and Guadalquivir River systems and mixing with recycled detritus from Cenozoic sediments and sedimentary rocks produces highly mature sands along the coast of the Huelva Province (SW Spain; SP46, SP45, SP44, SP43). Sands near the mouth of the Guadiana River (SP46) contain abundant quartz and minor feldspar and lithic grains. If these sands were to be melted with pure trona as a flux, the resulting glasses would contain very high SiO_2 values. The aluminosilicates in the sand would provide enough Al_2O_3. The sands located near the Guadalquivir mouth and further from the river mouths are even more mature (SP45, SP44, SP43). They contain very small amounts of feldspar and lithic fragments resulting in even higher SiO_2 and low Al_2O_3 in the calculated glasses. This suggests that abrasion by wave action is very efficient in breaking down these relatively unstable grains. Limestone or shell fragments are also rather scarce, making the CaO concentration in the glass too low. This could be resolved by blending additional $CaCO_3$ with the glass batch. In that way, sands from close to the Guadiana River mouth (SP46) would produce typical Roman natron glass. Glass produced with sand found more in the centre of the Gulf of Cádiz or near the Guadalquivir River and extra lime would be characteristically low in Al_2O_3. Sand from near the city of Huelva (SP45) would produce a glass with all major and minor elements similar to typical Roman natron glass, except for the low Al_2O_3 (1.5%). Further east (SP44) the Al_2O_3 (1.0%) content has dropped even more and also the P_2O_5 (0.018%) concentration would be lower than that of an average natron glass. Minor additional feldspar provided by the Guadalquivir (SP43) results in slightly higher Al_2O_3 (1.8%) in sands near the river mouth.

2.2.3   Volturno sand as a raw material for Roman glassmaking

The Volturno River sands were specifically mentioned by Pliny the Elder in the context of glass production. However, elemental analyses presented in this and other studies (Brill, 1999; Vallotto and Verità, 2002; Silvestri et al., 2006) show that sands from the coastal area between Cuma and Liternum are not suitable to produce glass. There is some variability among analysed sands from this region, but generally they are just insufficient in SiO_2 and contain too much Al_2O_3, CaO, Fe_2O_3 and K_2O due to the abundant augite, feldspar and volcanic lithic fragments in the sand. Given the close relationship between the composition of beach sands and the local geology (e.g., Rizzini, 1974; Le Pera and Critelli, 1997; Garzanti et al., 2002, 2004; Critelli et al., 2003), it is very unlikely that the Campanian sands have changed significantly during the past 2000 years. The Pleistocene–Holocene volcanic rocks of the Roccamonfina and Somma–Vesuvius will have left their mark on the composition of the beach sands for at least 400,000 years (Brocchini et al., 2001; Giannetti, 2001). Therefore, it must be concluded that the Volturno sands were probably not used to produce glass and that the writings of Pliny the Elder on this aspect are incorrect.

2.2.4 Background levels of manganese in sand raw materials

When interpreting the presented data, a few other interesting remarks can be made. Firstly, the MnO concentrations of the calculated glasses all fall within the ranges of Roman glass. This is due to the fact that MnO was often added to the glass as a decolourising agent (Sayre and Smith, 1961; Sayre, 1963; Henderson, 1985; Jackson, 2005; Freestone, 2008; Silvestri, 2008). In analytical studies of Roman glass, MnO is said to have been added deliberately when it occurs at levels above $0.1 - 0.2\%$ (Wedepohl et al., 2011a), 0.2% (Sayre, 1963), 0.4% (Brill, 1988), 0.5% (Jackson, 2005) or 1% (Henderson, 1985; Mirti et al., 2000, 2001). Our results indicate that background levels of MnO, i.e., the amount of MnO coming in with the sand raw material, are even lower. The highest MnO concentration that was measured resulted in 0.29% MnO in the calculated glass (sample IT74 from Pellaro in the Calabria Region, S Italy). For the sands which are actually suitable for Roman glass production, MnO concentrations vary between 0.004 and 0.078% (concentration in the calculated glass). We therefore suggest that the maximum amount of MnO that can be attributed to impurities in the sand raw material is lower than 0.1%. The levels at which the MnO actually works as a decolouriser depends on the amount of iron present in the glass. Given the average concentration of Fe_2O_3 in Roman glass (0.62%; Foster and Jackson, 2009) and the fact noted by Silvestri et al. (2005, 2008) that the decolouring effect is only effective when the MnO/Fe_2O_3 ratio is higher than 2, it seems that at least 1% of MnO was necessary to be sure that the glass would be decolourised. Concentrations of MnO that lie between 0.1 and 1% have probably been influenced by recycling of glass cullet.

It has been previously suggested that Volturno River sands contain elevated MnO contents and that this might be the reason why this sand would have been used to produce colourless glass (Baxter et al., 2005; Jackson, 2005). In support of this statement, reference is made to the analyses of Volturno sand reported by Brill (1999, 475). The sample in question (sample 4554), however, is coarse black sand with very high amounts of diopside, augite and hydrogrossular (Degryse and Schneider, 2008) and contains, in addition to the 0.43% MnO, 31.4% Fe_2O_3, 6.84% Al_2O_3, 8.87% MgO and only 33.70% SiO_2, making it highly unsuitable for glass production. The MnO contents reported in other Volturno sands are much lower, i.e., between 0.01 and 0.16% (Brill, 1999; Vallotto and Verità, 2002; Silvestri et al., 2006; samples IT23 to IT28 in this study). This is definitely too low to have a noticeable decolourising effect.

2.2.5   Soda introduced by the sand raw material

Another interesting observation is that the sand raw materials themselves bring part of the Na_2O to the glass. Glass produced from sand sample IT75 (from Bova Marina in the Calabria Region, S Italy) would contain 2.94% of sand-derived Na_2O. This is 17.7% of the total amount of Na_2O in the calculated glass and the maximum found in the present study. However, sands which contain high Na_2O are also high in Al_2O_3 indicating that the sodium is mostly derived from plagioclase. Suitable glassmaking sands generally contain less Na_2O. The 12 suitable sands discussed above generally bring between 0.16 and 0.96% of Na_2O (or 0.96 to 5.68% of the total) to the glass. Only sand sample FR17 contains significantly more Na_2O (1.96% of sand-derived Na_2O, 11.80% of the total Na_2O in the glass). Correspondingly, this sample also has an elevated Al_2O_3 (6.1%) level. This possible important contribution of soda coming from the sand should be considered when discussing 'glass batch recipes' and the proportion of sand to flux (Silvestri et al., 2006; Smedley et al., 1998; Smedley and Jackson, 2002), or when trying to calculate the composition of possible sand raw materials starting from the composition of the glass (e.g., Vallotto and Verità, 2002).

2.3   Melting experiments

In order to check our glass calculations and verify the conclusions drawn from them concerning the suitability of the sands, a number of glass melting experiments were performed. Glass was made using sands which were thought to be suitable for Roman natron glass production, either with or without the need for additional lime, based on the calculated glass compositions. Another melting experiment was performed using sand unsuitable for the production of Roman glass to check whether this would lead to the formation of a cotectic melt with a residual crystalline buffer (Rehren, 2000; Shugar and Rehren, 2002).

Batches were prepared using calculated amounts of natural beach sands combined with synthetic sodium minerals to produce glass with a Na_2O concentration equal to that of average Roman glass (i.e., 16.63%; Foster and Jackson, 2009). One of the melts was made with lime deficient sand. This was compensated for by the addition of extra $CaCO_3$ in the form of shell fragments. Modern seashells were collected from several beaches in southern France and northwest Italy, washed to remove any adhered sand grains, oven dried and crushed in the lab. The full analytical procedure used, is described in Brems (2012).

2.3.1   Experimental melting of suitable sands

Experimental melts were made using each of the three sands suitable to produce glass with a typical Roman composition without the need for extra calcium carbonate (Table 2.2). All three batches produced clear transparent glass. Glass made with sand IT85 had a green colour, while glasses IT87 and IT34 were pale green (Fig. 2.7a). All glasses contained small gas bubbles and dark dots which appear to be unmelted dark minerals (Fig. 2.7b). Microscopic analysis shows that these grains are mainly pyroxenes and opaque minerals. Whereas these minerals have an angular shape in the original sand raw materials, in the glass they are very well rounded and show a much smaller size distribution. This indicates that the heavy minerals in the sand raw material were at least partly absorbed by the glass and suggests that if the melting period were extended they would get completely dissolved.

Sample	Sand (g)	Na_2CO_3 (g)	Shell (g)	Total batch (g)	Batch - LOI (g) (Expected glass)	Final glass (g)	Colour
IT85	768,2	231,8	/	1000,0	844,6	448,5	Green
IT87	458,5	141,5	/	600,0	506,8	507,4	Pale green
IT34	459,9	140,1	/	600,0	512,1	483,3	Pale green
FR16	398,1	137,0	64,9	600,0	508,2	506,7	Blue
IT45	469,9	130,1	/	600,0	486,1	487,4	Brown frit

Table 2.2:
Calculated glass batches in order to produce glass with 16.63% Na_2O.

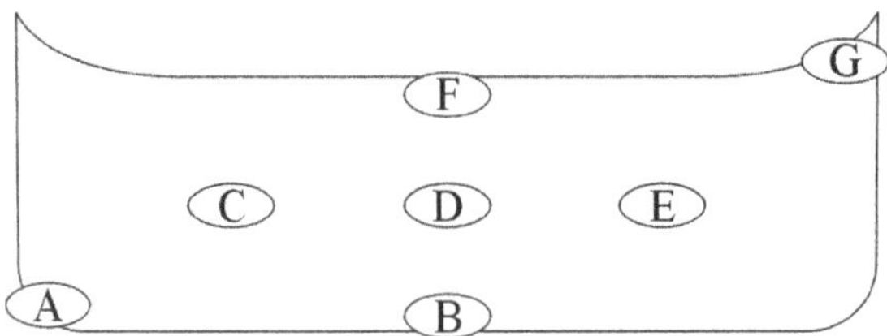

Figure 2.6:
Schematic cross section through the experimental glasses with indication of the sampling locations.

The major and minor element compositions of the analysed glass samples produced with sands IT85, IT87 and IT34 are presented in Table 2.3 and Fig. 2.8. The measured concentrations generally lie close to the expected values, although there are important heterogeneities within the glasses. The upper part of the glass appears to be enriched in SiO_2, while the lower part of the glass is relatively higher in CaO and Na$_2$O. A similar spatial variability for the three major base glass components has been reported before and could be explained either by a separation of liquids with different densities during the initial stages of the melting process or by the movement of undissolved grains in the melt (Tiede and Tooley, 1945; Cable and Bower, 1965; Chopinet et al., 2010). According to the first possible explanation, locally produced silica-rich liquids migrate upwards and CaO-rich liquids downwards under the influence of gravity. The second explanation suggests that undissolved quartz grains would float to the top of the crucible during glass melting, while fragments of free lime, resulting from decarbonised CaCO$_3$ grains, would sink to the bottom of the crucible.

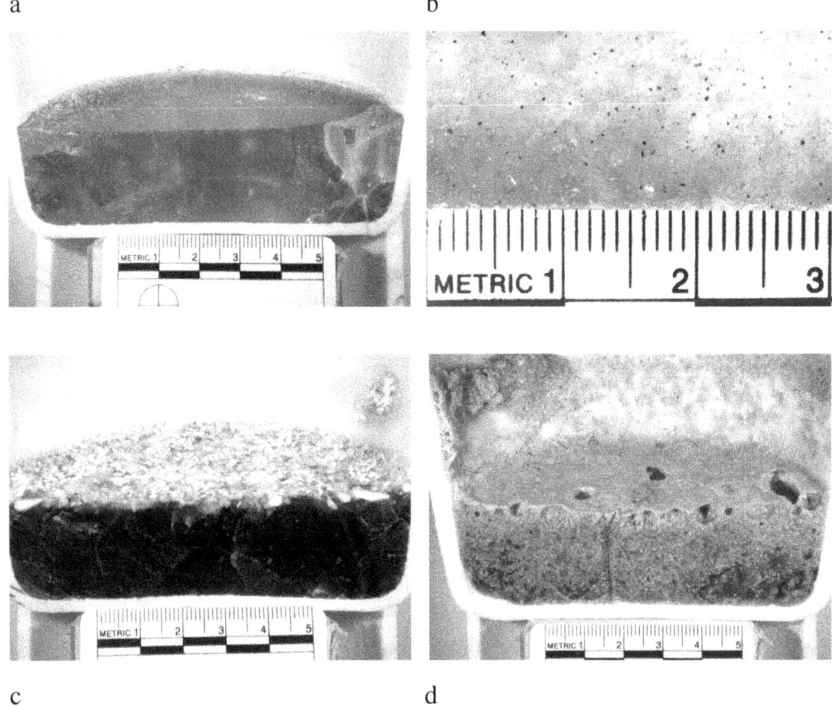

Figure 2.7:
(a) Pale green glass produced with sand IT34. The glass contains small gas bubbles and unmelted heavy minerals. (b) Detail of the lower surface of glass IT87 showing the unmelted heavy minerals (dominantly pyroxenes). (c) Blue glass produced with sand FR16 and extra shell fragments. Inclusions of powdery white free lime occur over almost the entire surface of the glass. (d) Brown frit produced with sand IT45. This glass batch was only partly melted and contains abundant unmelted and newly formed crystalline phases embedded in a glass matrix. Reprinted from Brems et al. 2012a, Glass Technology: European Journal of Glass Science and Technology A.

Table 2.3:

	IT85 (cal)	IT85.B	IT85.A	IT85.C	IT85.D	IT85.E	IT85.F	IT85.G	IT87 (cal)	IT87.C	IT87.E	IT34 (cal)	IT34.B	IT34.A	IT34.D	IT34.F	IT34.G
SiO_2	69,57	66,25	67,51	67,43	67,18	67,42	68,85	69,96	72,61	70,97	71,39	71,42	69,94	70,42	71,49	72,11	73,15
Al_2O_3	2,43	2,88	2,74	2,64	2,62	2,64	3,31	3,82	1,14	1,36	1,40	2,55	2,74	3,06	2,90	3,09	3,52
$Fe_2O_3(t)$	1,32	1,66	1,55	1,52	1,53	1,53	1,36	1,26	0,49	0,86	0,58	0,89	0,79	0,78	0,77	0,73	0,75
MgO	0,53	0,82	0,79	0,77	0,77	0,77	0,68	0,63	0,28	0,36	0,35	0,31	0,36	0,35	0,35	0,33	0,33
MnO	0,08	0,11	0,11	0,11	0,11	0,11	0,09	0,08	0,02	0,03	0,03	0,07	0,08	0,08	0,08	0,07	0,07
CaO	8,49	10,57	10,58	9,94	9,93	9,84	8,56	7,91	8,13	8,55	8,53	7,18	8,03	7,76	7,42	7,08	6,83
Na_2O	16,63	16,62	15,65	16,53	16,77	16,60	16,06	15,26	16,63	17,08	16,92	16,63	17,07	16,58	16,02	15,60	14,38
K_2O	0,81	0,74	0,76	0,75	0,75	0,75	0,79	0,79	0,63	0,70	0,70	0,86	0,85	0,85	0,87	0,86	0,86
TiO_2	0,10	0,30	0,28	0,28	0,28	0,28	0,25	0,23	0,04	0,07	0,07	0,07	0,11	0,10	0,10	0,09	0,09
P_2O_5	0,04	0,04	0,04	0,04	0,05	0,05	0,05	0,06	0,02	0,02	0,02	0,02	0,02	0,02	0,02	0,02	0,02
LOI	0,00	0,00	0,00	0,00	0,00	0,00	0,00	0,00	0,00	0,00	0,00	0,00	0,00	0,00	0,00	0,00	0,00
Total	100,00	100,00	100,00	100,00	100,00	100,00	100,00	100,00	100,00	100,00	100,00	100,00	100,00	100,00	100,00	100,00	100,00

	FR16 (cal)	FR16.B	FR16.A	FR16.D	FR16.F	FR16.G	FR162.A	FR162.B	FR163.A	FR163.B	IT45 (cal)	IT45.B	IT45.F	Shell	Na_2CO_3	IT34.G
SiO_2	70,21	69,24	69,53	69,39	72,18	73,51	70,06	69,87	69,93	69,77	48,66	48,25	47,97	0,68	0,29	73,15
Al_2O_3	3,29	3,46	3,56	3,35	3,25	4,15	3,46	3,49	3,43	3,45	5,21	5,45	5,17	0,04	0,00	3,52
$Fe_2O_3(t)$	1,28	1,29	1,27	1,35	1,19	1,24	1,32	1,29	1,30	1,31	3,49	4,55	4,16	0,01	0,00	0,75
MgO	0,21	0,25	0,27	0,26	0,23	0,23	0,25	0,25	0,25	0,25	4,37	4,50	4,87	0,06	0,01	0,33
MnO	0,01	0,01	0,01	0,01	0,01	0,01	0,01	0,01	0,01	0,01	0,09	0,10	0,10	0,00	0,00	0,07
CaO	7,48	8,07	8,15	7,55	6,64	6,36	7,72	7,78	7,66	7,54	19,53	18,67	20,42	53,85	0,00	6,83
Na_2O	16,63	16,79	16,29	17,16	15,64	13,66	16,20	16,33	16,39	16,67	16,63	16,50	15,50	0,72	45,74	14,38
K_2O	0,63	0,63	0,64	0,67	0,61	0,58	0,73	0,73	0,77	0,74	1,56	1,44	1,33	0,03	0,00	0,86
TiO_2	0,23	0,22	0,23	0,23	0,21	0,22	0,22	0,22	0,22	0,22	0,38	0,45	0,42	0,00	0,00	0,09
P_2O_5	0,03	0,03	0,03	0,03	0,03	0,03	0,03	0,03	0,03	0,03	0,09	0,09	0,08	0,02	0,01	0,02
LOI	0,00	0,00	0,00	0,00	0,00	0,00	0,00	0,00	0,00	0,00	0,00	0,00	0,00	44,59	53,95	0,00
Total	100,00	100,00	100,00	100,00	100,00	100,00	100,00	100,00	100,00	100,00	100,00	100,00	100,00	100,00	100,00	100,00

Elemental composition of the experimental glass samples, shell fragments and Na_2CO_3 as determined by ICP–OES analysis. All results are in wt%. The calculated glass compositions for the sands used in the melting experiments are given for comparison (cal).

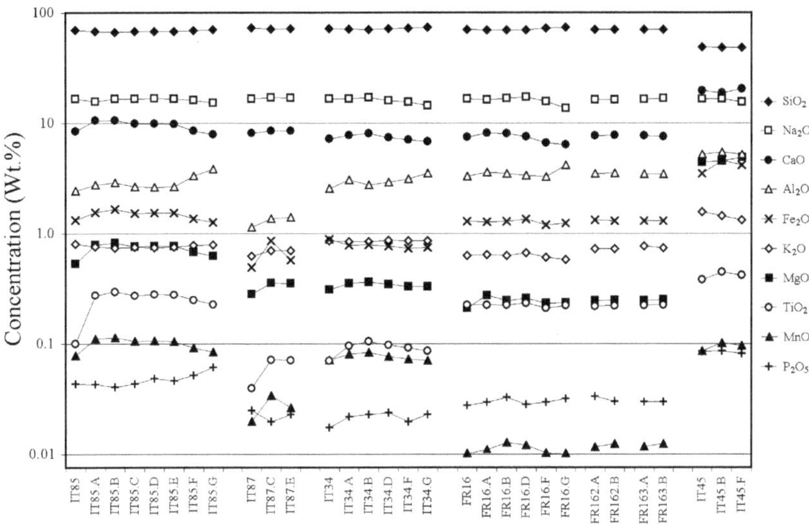

Figure 2.8:
Comparison between the calculated glass compositions (IT85, IT87, IT34, FR16 and IT45) and the compositions of the experimental glasses. For the location of the samples analysed within the volume of glass, see Fig. 2.6

Our results suggest there are also some depth dependent variations in the minor elements in the glass. The glass at the upper part of the crucible is enriched in Al_2O_3 and K_2O. This can be explained by the flotation of undissolved feldspar grains which have similar densities as quartz. The lower part of the glass has higher concentrations of Fe_2O_3, MgO, TiO_2 and MnO. These are probably related to the heavy mineral fraction of the sand which would have the tendency to sink in the molten glass before dissolving.

All of the glass samples analysed are higher in Al_2O_3 than the calculated glass (Table 2.3 and Fig. 2.8). This indicates that the molten glass reacted with the walls of the Al_2O_3 crucible and extracted about 0.5% Al_2O_3 from them. A similar increase in Al_2O_3 was also observed by Gerth et al. (1998) in their melting experiments using Al_2O_3 crucibles.

2.3.2   Experimental melting of lime deficient sand

Sand FR16 contained too little CaO to produce a stable glass. Therefore, extra shell fragments were added to the batch (Table 2.2). The resulting glass was clear transparent, had a blue colour and contained small gas bubbles and small dark mineral crystals similar to the previously described glasses. However, with this batch the melting process was not complete. In the upper 5 mm of the glass and almost over its entire surface, white powdery masses of up to 2 mm in size occurred (Fig. 2.7c). The mineralogical composition of this material was analysed by XRD, which showed the presence of free lime (CaO), portlandite ($Ca(OH)_2$) and minor amounts of wollastonite ($CaSiO_3$) and quartz (SiO_2). So the remaining material originated from the shell fragments in the batch which had not fully reacted with the rest of the raw materials. No calcite or aragonite was detected, indicating that the shell fragments did completely decompose to free lime. Due to the absorption of water from the air during and after cooling of the glass, part of the free lime was transformed into portlandite. After longer contact with the air, all free lime would react to form portlandite and eventually calcium carbonate (Paynter and Dungworth, 2011).

This concentration of undissolved lime particles at the upper surface of our experimental glass seems to be in contradiction with the previously cited explanation for the compositional heterogeneity of glass: because of its high density, free lime is expected to sink in the molten glass (Cable and Bower, 1965; Chopinet et al., 2010). One possible explanation for this is that the CaO grains were kept afloat in the glass by gas bubbles. However, the results of the chemical analyses show that the upper part of the glass (with the lime inclusions) is not enriched in CaO (Table 2.3 and Fig. 2.8). On the contrary, we find the same depth dependent chemical variations as in the glasses made without the addition of extra shell fragments: the upper part of the glass is enriched in SiO_2, while the lower part is higher in CaO and Na_2O. This might indicate that the movement of relatively heavy CaO fragments under the influence of gravity is only a minor process and certainly does not affect all free lime grains in the batch. This is very well plausible since the speed at which a grain can settle in a liquid is dependent on a wide number of factors such as the viscosity of the liquid, the difference in density between the liquid and the grain, the shape and size of the grain, and the presence of gas bubbles adhered to the grain.

As observed in the previous melting experiments, the average Al_2O_3 content of the produced glass was found to be slightly higher than expected (Table 2.3 and Fig. 2.8), indicating that additional Al_2O_3 was extracted from the crucible walls.

In order to check whether remelting of the resulting glass and lime phases would result in the formation of a homogeneous glass, the material was broken up and remelted. For these experiments, smaller mullite crucibles were used. In a first crucible a number of chunks, measuring between 5 and 25 mm, of the blue glass with inclusions of unreacted lime were remelted together. For a second batch, the glass and the included lime phases were first crushed until a fraction smaller than 1 mm was obtained.

The remelting of the finely crushed glass resulted in the formation of a homogeneous, blue transparent glass. This glass no longer contained dark mineral inclusions and only minor gas bubbles. The glass produced using the larger glass chunks still contained a number of white powdery lime inclusions up to 2 mm in size. These inclusions were again concentrated in the upper 5 mm of the glass. The glass phase itself was visually identical to that produced by remelting the finer crushed glass. The chemical similarity of the two glasses is shown by the ICP–OES results (Table 2.3 and Fig. 2.8). For each of the two batches, two fragments of the central part of the glass were analysed. Both glasses were found to be indistinguishable within analytical error with compositions very close to the calculated glass composition for all elements measured.

These experiments indicate that if extra pieces of shell or limestone were added to the glass batch, these would have to be finely crushed. If not, unreacted lime would remain in the glass. A prolongation of the applied melting time would probably also have a positive effect on the degree of absorption of the lime phases and the final homogeneity of the glass.

2.3.3   Experimental partial melting of unsuitable sand

In a final melting experiment we wanted to try to produce a cotectic glass melt using a sand raw material with a theoretical melting temperature higher than the applied firing temperature according to Rehren (2000) and Shugar and Rehren (2002). To select an appropriate sand for this experiment, we reduced the elemental compositions of our calculated hypothetical glasses to their three most basic oxides (SiO_2, Na_2O and CaO) and plotted them on the soda–lime–silica ternary phase diagram (Shahid and Glasser, 1972) using the method described by Rehren (2000) (Fig. 2.9). For the cotectic melting experiment we chose sand sample IT45 with a theoretical melting temperature of approximately 1200°C. This sand came from the beach just to the north of the Garigliano River mouth in the most southern part of the Lazio Region (W Italy). According to our criteria used to evaluate the sand raw materials, this sand is not suitable to produce Roman natron glass. It is mainly composed of quartz, calcite, feldspar and augite, with minor dolomite, aragonite and garnet. It contains insufficient SiO_2 and elevated CaO, Al_2O_3, Fe_2O_3, MgO and K_2O levels. The hypothetical glass that would be produced if sand IT45 was completely melted after the addition of pure natron would have the following composition: 48.7% SiO_2, 16.6% Na_2O, 19.5% CaO, 5.2% Al_2O_3, 3.5% Fe_2O_3, 4.4% MgO, 1.6% K_2O, 0.09% MnO, 0.38% TiO_2 and 0.09% P_2O_5.

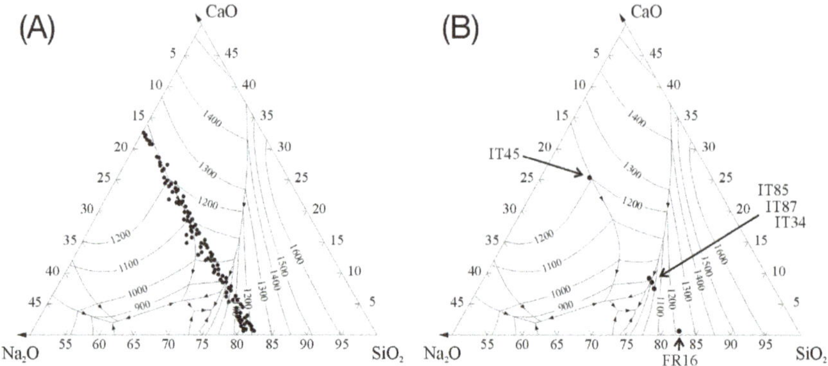

Figure 2.9:
Na$_2$O-CaO-SiO$_2$ ternary phase diagrams (Shahid and Glasser, 1972) showing the composition of the calculated glasses based on the analysed beach sands. The data are reduced to fit the diagram as explained in the text. (a) All hypothetical glasses made with the addition of pure natron. (b) Hypothetical glasses made with the sands used in the melting experiments and pure natron. Reprinted from Brems et al. 2012a, Glass Technology: European Journal of Glass Science and Technology A.

After firing, the batch had only partly melted, producing a brown frit (Fig. 2.7d). Most of the raw materials recrystallised to form new mineral phases which are dominated by combeite (Na$_2$Ca$_2$Si$_3$O$_9$) and melilite ((Ca,Na)$_2$(Al,Mg,Fe)(Si,Al)$_2$O$_7$). The XRD patterns also show the presence of newly formed nepheline, a sodium-rich feldspathoid. Unreacted raw materials include quartz grains, feldspar and heavy minerals, mostly augite. These newly formed and unmelted minerals are cemented together by a glass matrix. Locally, small pockets of glass, up to 3 mm in size, occur between the residual crystalline matter. Near the upper surface large gas bubbles (up to 15 mm) occur. The chemical composition of the produced frit lies very close to the calculated values (Table 2.3 and Fig. 2.8).

Shugar and Rehren (2002) performed a series of experimental melts using chemical-grade raw materials (i.e., pure silica, calcium carbonate and sodium carbonate) to check their partial melting theory. They reported that batches heated to temperatures lower than the theoretical melting temperature showed evidence of partial melting with glassy sections forming under a crystalline cap containing the rest of the partially reacted batch material. After crushing, they were able to manually separate a clean glass fraction from the residual crystalline matter if about half the total material appeared glassy after firing. However, they also indicated that if the difference between the theoretical batch melting temperature and the actual firing temperature was a bit larger, the amount of glass produced was rather low. This can also be concluded from the experiments performed in this study and indicates that, even with a process such as partial melting in play, in order to produce a significant amount of glass, the composition of the entire batch must lie close to the composition of average Roman glass (i.e., the eutectic region in the soda–lime–silica phase diagram). Therefore, the elemental compositional range of possible sand raw materials would still be rather limited.

2.3.4 Batch mixtures: ratio of sand to natron

The glass calculations and melting experiments performed in this study can provide some insights in the glass batch recipes that were used by the Roman glassmakers. The few written accounts of ancient glassmaking state that the sand and alkali raw materials were added to the batch in 'parts' (Smedley et al., 1998; Smedley and Jackson, 2002). However, these sources do not specify whether these parts were by weight or volume. The majority of medieval publications quote a ratio of 1:2 sand to wood ash for glassmaking (Smedley and Jackson, 2002; Jackson and Smedley, 2004) and after performing some batch calculations and melting experiments, Smedley et al. (1998) and Smedley and Jackson (2002) concluded that these parts are a measurement by weight. In the case of Roman natron glass, most translations of Pliny the Elder's Natural History state that one part of sand was to be mixed with three parts of natron and that these parts could be measured either by weight or volume (Bostock and Riley, 1857; Turner, 1956; Smedley and Jackson, 2002; Silvestri et al., 2006). However, Rottländer (1986) and Healy (1999) argue that this sentence was wrongly translated and that Pliny in fact meant that one part of sand must be mixed with three quarters of natron.

In our melting experiments we used calculated amounts of sand and synthetic sodium carbonate to produces glass with a Na_2O concentration of 16.63% (Foster and Jackson, 2009). This led to batch recipes with a Na_2CO_3 to sand ratio of 0.30, measured by weight. When we calculate the Na_2CO_3 back to pure trona $(Na_2CO_3 \cdot NaHCO_3 \cdot 2H_2O)$, the trona/sand ratio would become 0.43 by weight. Depending on the amount of nonreactive salts (chlorides and sulphates) and structural water in the available natron, this ratio would of course increase. If the natron raw material contains 10% of these constituents, the natron/sand ratio would have to be 0.48 by weight in order to attain enough Na_2O in the glass. If this nonreactive part rises to 20% of the natron, the natron/sand ratio would be 0.54. When looking again at the two batch recipes that were suggested based on the different translations of Pliny the Elder's Natural History, we can calculate that if a batch recipe was used where one part of sand was mixed with three quarters of natron, as suggested by Rottländer (1986) and Healy (1999), the nonreactive salts and water would account for 43% of the used natron. If a batch recipe with a natron over sand ratio of three was used to produce Roman glass, the excess salts would account for 86% of the used natron. These calculations suggest that the former batch recipe would be the more likely one.

The possibility that the ratio mentioned in Pliny's recipe had to be measured by volume, rather than by weight, is more difficult to evaluate. Brill (1988) calculated that Roman natron glass from Jalame consisted of a mixture of Belus sand and Egyptian natron in proportions of approximately 5 to 2 parts by weight (i.e. a natron/sand ratio of 0.40). This ratio is very similar to the one we found for glass produced with pure trona. After estimating the bulk densities of Belus River sand (1.44 g/ml) and crushed natron (1.24 g/ml), Brill (1988) was able to show that very similar glass could be produced following a recipe of two parts of sand to one part of natron, measured by volume. This might have been a more convenient way of measuring ingredients for the Roman glass maker (Brill, 1988). However, since the weight of a standard volume depends greatly on factors such as grain size and packing (Smedley and Jackson, 2002), a recipe measured by volume might not always produce glasses with the same composition and as a result different working properties.

2.4   Conclusions

In this chapter, we calculated the composition of hypothetical glasses that can be made from modern beach sands from Spain, France and Italy, and we performed a series of glass melting experiments to reproduce Roman natron glass. The results of our calculations and experiments show that Roman-type glass making sands are relatively rare. Six limited areas could be defined where suitable sand raw materials would have been available to the Roman glassmaker, either with or without the need for an additional source of lime (Fig. 2.10):

(1) Beach sand between the mouth of the River Basento and the River Bradano (Basilicata Region, SE Italy) is suitable for glass production in its current state and would produce a typical Roman natron glass.
(2) In the area southeast of Brindisi, on the northeastern side of the Salentina peninsula (Puglia Region, SE Italy), beach sands would produce glass with chemical compositions within the ranges of Roman glass for almost all major and minor elements. Only the Al_2O_3 concentration is lower than that of typical natron glass.

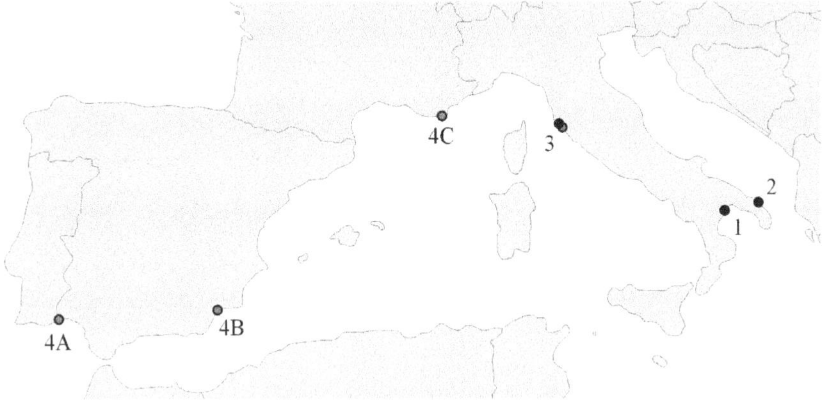

Figure 2.10:
Map of the western Mediterranean showing the areas where suitable sand raw materials occur. Black: sand suitable without the need for extra $CaCO_3$. Grey: sand suitable after the addition of extra $CaCO_3$. See text for description of the locations. Reprinted from Journal of Archaeological Science, Vol 39, Brems, D., Degryse, P., Hasendoncks, F., Gimeno, D., Silvestri, A., Vassilieva, E., Luypaers, S., Honings, J., Western Mediterranean sand deposits as a raw material for Roman glass production, 2897-2907, Copyright (2012), with permission from Elsevier.

(3) Sands from the western part of the Follonica Gulf, between Piombino and Follonica (Tuscany Region), are suitable for glass production. Except for the low P_2O_5 concentration, all major and minor elements analysed fall within the range of Roman glass. In the eastern part of the gulf, between Follonica and Punta Ala, sands contain less shell, resulting in lower CaO contents. These sands are only suitable after the addition of extra calcium carbonate.

(4) Beach sands (a) near the mouth of the Rio Guadiana (Huelva Province, SW Spain), (b) near the town of Águilas in the Murcia Region (SE Spain), and (c) from the Bay of Hyère (Département du Var, Provence, SE France) are all very rich in quartz and contain only very small amounts of shell or limestone fragments. After the addition of extra $CaCO_3$ however, these sands would produce glasses with a typical Roman composition.

These results of course do not prove that there was a Roman primary glass production industry in the western Mediterranean, but it demonstrates that it would have been possible and that if it existed is was most likely in one of the suggested regions.

Chapter 3

The Sr-Nd isotopic fingerprint of sand raw materials

D. Brems, M. Ganio, P. Degryse

In the previous chapter, we have evaluated beach sands from the western part of the Mediterranean (Spain, France and Italy) for their suitability for Roman natron glass production. It was shown that Roman-type glassmaking sands only occur in a few restricted areas. Now we must find a way to distinguish glass that could have been made using sand from these sources from each other and from the known glass production centres in Egypt and Syro-Palestine. However, over the past decades it has proven very difficult to link a glass artefact to a particular source or site of production (e.g. Degryse et al., 2009a). Since the major and minor elemental composition of Roman natron glasses was found to be relatively uniform (Freestone, 2006; Wedepohl et al., 2011a) and almost never diagnostic for their origin, other methods had to be found to provenance ancient glass artefacts. In recent years, the use of trace elements and radiogenic isotopes were shown to offer great potential. Radiogenic isotopes are often used in earth sciences and sedimentary geology, for a wide range of applications, such as dating the formation of rocks and minerals (geochronology), chemical stratigraphic correlation, and tracing the sources and transport of dissolved and detrital constituents in sedimentary, hydrologic and biogeochemical cycles (Banner, 2004). Particularly Sr and Nd isotope ratios are powerful tracers for sediments. Because of their relatively large masses and small relative mass differences, the isotopic fractionation of the different isotopes of Sr and Nd is negligible in this context (Stille and Shields, 1997; Banner, 2004). Therefore, the isotopic composition of these elements in a glass is believed to be identical to that of the raw materials from which it was derived.

More and more isotopic data of ancient glass is being published. However, until now there was no database of isotopic signatures of possible sand raw materials available for comparison. In this chapter, we present such a database of $^{87}Sr/^{86}Sr$ and $^{143}Nd/^{144}Nd$ isotope ratios of beach sand deposits from the western part of the Mediterranean area, some of which were identified as suitable raw materials for Roman natron glass production.

3.1   The origin of Sr in ancient natron glass

Strontium is a trace element which readily substitutes for calcium in minerals, such as aragonite, calcite and plagioclase. Differences in Sr isotopic signatures of different types of rock allows them to be used as a tracer for atmospheric dust and detrital components in sedimentary basins (Stanley et al., 2003; Banner, 2004; Grousset and Biscaye, 2005). Regional variations in $^{87}Sr/^{86}Sr$ values of sediments are a function of the following features of the source rocks: (1) the mineralogy, age, and crustal versus mantle source for igneous and metamorphic rocks; (2) the provenance and maturity for sandstones and shales; and (3) the age and extent of alteration for marine carbonates, evaporites and phosphorites (Wedepohl, 1978; Banner, 2004).

Because of a large difference in distribution coefficients for strontium in the different calcium carbonate minerals, the Sr content of seashells and limestones is controlled by the mineralogy. In equilibrium with seawater, aragonite and calcite contain about 8,000 and 1,200 ppm Sr, respectively (Kinsman, 1969; Katz et al., 1972). Intermediate values are due to mixed mineralogy of the specimens (Wedepohl, 1978). Since aragonite and high-Mg calcite are metastable, they are transformed into stable low-Mg calcite during diagenesis. During this dissolution–precipitation process, the newly precipitated low-Mg calcite will incorporate less Sr into its crystal lattice than the dissolving metastable phases. As a consequence, older limestones with stabilised mineralogy have Sr concentrations an order of magnitude lower than those in the originally deposited material (around 400 ppm; Kinsman, 1969; Katz et al., 1972; Veizer, 1977; Wedepohl, 1978).

In calcareous sediments, such as the beach sands suited for the production of natron glass, strontium is expected to be mainly contained in the carbonate fractions. $^{87}Sr/^{86}Sr$ ratios of the carbonate are the same as those of the seawater in which it formed and because of the high Sr content, a small amount of calcium carbonate will usually mask the isotopic signature of the silicate fraction (Wedepohl, 1978). Depending on the mineralogy and the amount of carbonate present in the sand, varying influences on the bulk $^{87}Sr/^{86}Sr$ isotope ratios may be attributed to the detrital silicate phases (mainly feldspar or clays). The average Sr concentrations in sands are in the 30 – 400 ppm range (Wedepohl, 1978).

Roman natron glass contains between 5 and 10% CaO (Foster and Jackson, 2009). The bulk of the Sr in Roman glass is believed to have been incorporated with the lime-bearing material (Wedepohl and Baumann, 2000; Freestone et al., 2003). Where the lime was derived from Holocene seashell, the Sr isotopic composition of the glass is similar to that of modern-day seawater. Where the lime was derived from 'geologically aged' limestone, the Sr isotopic signature reflects that of seawater at the time the limestone was deposited, possibly modified by diagenesis. However, other minerals in the sand, such as feldspar and mica, can also influence the Sr budget of the glass batch (Freestone et al., 2003; Degryse et al., 2006a). The contribution of natron to the Sr content and Sr isotopic signature of glass is negligible (Freestone et al., 2003). The Sr content of natron glass is also a useful indicator of the source of lime. Since aragonitic seashell may contain a few thousand ppm Sr and calcitic limestone will only incorporate a few hundred ppm of Sr, a similar difference in Sr concentration can be expected in glass produced with these two different sources of lime (Wedepohl and Baumann, 2000; Freestone et al., 2003). Natron glass melted using limestone contains less than 200 ppm Sr while shell fragments can bring 300 to 600 ppm Sr to the glass.

3.2   The Nd isotopic composition of ancient glass

Samarium and neodymium are both light rare earth elements (LREE) belonging to the lanthanide series. The application of variations in the isotopic composition of Nd to sediments is based on the fact that accumulated clastic sediments are basically just the mechanical disintegration products of igneous, metamorphic and older sedimentary rocks exposed in the source areas (Goldstein et al., 1984; DePaolo, 1988; Grousset et al., 1988; Jeandel et al., 2007, and references therein). The Nd isotopic signature of the source terrain is generally preserved in the resulting sediments. Consequently, variations in the isotopic composition of Nd are very useful as tracers in sediment provenance studies (Linn and DePaolo, 1993; Banner, 2004; Grousset and Biscaye, 2005).

Sm and Nd are enriched in accessory minerals, such as apatite, titanite, allanite, perovskite, xenotime and monazite (Wedepohl, 1978; Foster and Vance, 2006; McFarlane and McCulloch, 2007). They also occur in trace concentrations in many rock-forming minerals like feldspar, biotite, amphibole and clinopyroxene, in which they replace major ions (Best, 2003; Faure and Mensing, 2005). Quartz contains virtually no Sm or Nd. The concentration of Nd in siliciclastic sediments and sedimentary rocks is usually in the order of 5 – 50 ppm (Faure and Mensing, 2005). In limestone and shell, the absolute Nd content is even lower (between 0.5 and 10 ppm; Wedepohl, 1978; Faure and Mensing, 2005; Wedepohl et al., 2011b). Natron appears to contain hardly any Nd, i.e. in the order of 20 – 40 ppb Nd (Wedepohl et al., 2011b; Shortland, unpublished data), and has consequently no influence on the Nd budget and the Nd isotopic signature of the glass. The Nd in Hellenistic, Roman and Early Byzantine period glass (i.e. natron based glass) thus originates from the heavy or non-quartz mineral fraction of the silica raw material (Degryse and Schneider, 2008).

Due to the varying sediment influx from the Nile (fluvial), the Sahara (aeolian) and the European continent (fluvial), the Nd isotopic composition of deep-sea sediments in the eastern Mediterranean Sea varies significantly. The River Nile has an exceptionally high ε_{Nd} value in its sediment load of around -1 (Weldeab et al., 2002; Scrivner et al., 2004), as it is dominated by young volcanic rocks from the Ethiopian Plateau. Sediments dominated by input from wind-blown Saharan dusts, on the other hand, show typical low (old) ε_{Nd} values of around -13 (Grousset et al., 1988, 1998; Henry et al., 1994). Sediments entering the Ionian Sea from the Calabrian Arc and the Adriatic Sea are characterised by low ε_{Nd} values of -11.06 (Weldeab et al., 2002). Aegean Sea sediments show average ε_{Nd} values of -7.89 (Weldeab et al., 2002). When these sediments enter the Mediterranean Sea, they are redistributed by the dominant sea currents (Pinardi and Masetti, 2000; Weldeab et al., 2002, Hamad et al., 2006). For example, Nile River sediments are transported eastwards along the Egyptian and Israeli coasts, possibly up to southern Turkey. Because of the combination of all these different sources and currents, the isotopic pattern of the eastern Mediterranean surface sediments shows a pronounced E–W gradient from as high as ε_{Nd} = -1 at the mouth of the River Nile and the coasts of Egypt and Israel to ε_{Nd} = -12 south of Sicily (Goldstein et al., 1984; Frost et al., 1986; Weldeab et al., 2002). Sediment samples from Alexandria (Egypt) show ε_{Nd} values between -8 and -6 (Freydier et al., 2001; Tachikawa et al., 2004). These values are significantly lower than the pure Nile end-member and suggest mixing between Nile particles and sediment with a Sahara origin coming from the west with the dominant sea currents.

In the western Mediterranean, the distribution of the Nd isotopic signatures is less well known. Only a few results for particulates from the Rhône and the Po Rivers and deep sea sediments near Gibraltar and the southern French coasts are published and they all show rather low signatures with ε_{Nd} between -10.8 and -9.7 (Frost et al., 1986; Grousset et al., 1988; Henry et al., 1994). One result from the Tyrrhenian Sea shows a ε_{Nd} value of -7.6 (Frost et al., 1986). Although the number of analyses is small, there seems to be a significant difference in Nd isotopic signatures between the easternmost part of the Mediterranean Sea and the rest of the basin. If the same regional variations in Nd isotopic signatures occur in sand deposits across the Mediterranean, this can be used to trace ancient glass artefacts to their primary origin.

3.3   The isotopic fingerprint of sand and primary glass the eastern Mediterranean

Beach sands from the eastern Mediterranean coasts near the mouth of the river Belus, were sampled by R.H. Brill in the sixties (Brill 1999) and analysed by Degryse and Schneider (2008) and Brems et al. (2012a, b). Supplemental sands were sampled in the framework of this project (see Appendix D). They show high Nd isotopic signatures between -1 and -4.8 ε_{Nd}, indicating the mixing of sands derived from the Nile with more local material, delivered by the Belus itself. Greek and Turkish sands have Nd values between -4.2 and -7.4 ε_{Nd} but were unsuitable for glass making due to their very low silica or too high alumina content. Primary glass from the factories at Bet Eli'ezer and Apollonia, using the Israeli coastal sand (Brill, 1999) and active between the 6th and 8th century AD, produce raw glass with an isotopic signature between -4.1 and -5.1 ε_{Nd} (see also Appendix D and Chapter 6). Freestone et al. (nd) earlier reported values for Bet Eli'ezer, Apollonia and HIMT primary glass between -5.0 and -6.0 ε_{Nd}. In contrast, inland Egyptian sands sampled along the Nile in the close western desert between Gizeh and Aswan, show Nd isotopic signatures between -6.0 and -8.7 ε_{Nd}, with a $^{87}Sr/^{86}Sr$ ratio between 0.7081 and 0.7115. Sands from the Sahara desert sampled in Egypt and Tunisia show Nd isotopic signatures between -10.0 and -14.2 ε_{Nd}, with a $^{87}Sr/^{86}Sr$ ratio between 0.7085 and 0.7210. Therefore, a Nd isotopic signature of ε_{Nd} higher than -7.0 seems to be the cut off for the primary origin of glass in the eastern Mediterranean, be it from Syro-Palestinian or Egyptian (coastal) factories (as suggested for HIMT glass; Nenna, 2014). It is clear, however, that more work is needed to construct an extensive database of Israeli and Egyptian sands to establish a firm background of the isotopic signatures of suitable glassmaking sands in this area.

3.4   The isotopic fingerprint of sand raw materials in the western Mediterranean

As described above, the apparent applicability of the Rb–Sr isotopic system to glass studies lies with the assumption that Sr is incorporated into the glass with the lime source. Because of the difference in Sr isotopic signatures of modern-day seashell and old limestone, the $^{87}Sr/^{86}Sr$ isotope ratio of the glass could be indicative for the nature of the lime source used. However, other minerals naturally included within the sand raw material can also contribute to the final Sr isotopic signature of the glass. The extent of this influence is not well understood.

Nd in ancient glass is incorporated with the source of silica. The variation in Nd isotopic signatures observed in deep sea sediments across the Mediterranean Sea offers great potential to distinguish possible sand raw materials and primary glass from the eastern and western part of the basin. However, it is of course not possible to directly compare the Nd isotopic signature of glass to that of sea-floor sediments. Sand deposits are often much more locally derived and it is not certain that these beach sands show the same regional variation in Nd isotopic composition.

In this chapter, we therefore investigated the variation in $^{87}Sr/^{86}Sr$ ratios of possible sand raw materials and the extent of the influence of Sr coming from the sand source on the final Sr isotopic signature of the resulting glass. We also studied whether variations in Nd isotopic signatures can distinguish sand deposits around the Mediterranean. 77 beach sand samples from Spain, France and Italy (Fig. 3.1 and Fig. 3.2) were analysed for their Sr and Nd isotopic compositions and it is assessed whether the regional pattern in Nd isotopic signatures of deep-sea sediments can be recognised in these beach sands. The suitability of the sands for Roman natron glass production has been evaluated in Chapter 2.

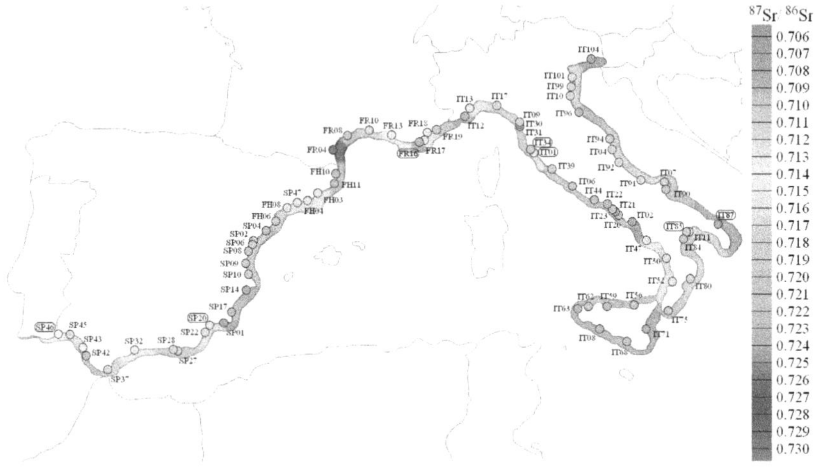

Figure 3.1:
Map of the western Mediterranean showing the sample locations and the $^{87}Sr/^{86}Sr$ isotope ratios of the beach sands analysed. Sand samples IT34, IT85 and IT87 were previously identified as good glassmaking sands. Sands SP46, SP20, FR16 and IT01 can be used to make Roman natron glass after the addition of extra lime to the glass batch. Reprinted from Archaeometry, Vol 55, Brems, D., Ganio, M., Latruwe, K., Balcaen, L., Carremans, M., Gimeno, D., Silvestri, A., Vanhaecke, F., Muchez, P., Degryse, P., Isotopes on the beach Part 1 – Strontium isotope ratios as a provenance indicator for lime raw materials used in Roman glassmaking, 214-234, Copyright (2013), with permission from John Wiley and Sons.

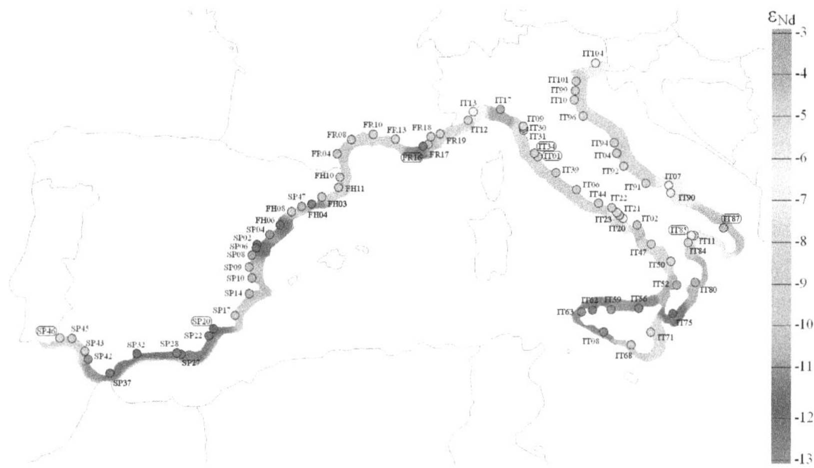

Figure 3.2:
Map of the western Mediterranean showing the sample locations and the ε_{Nd} values of the beach sands analysed. Sand samples IT34, IT85 and IT87 were previously identified as good glassmaking sands. Sands SP46, SP20, FR16 and IT01 can be used to make Roman natron glass after the addition of extra lime to the glass batch. Reprinted from Archaeometry, Vol 55, Brems, D., Ganio, M., Latruwe, K., Balcaen, L., Carremans, M., Gimeno, D., Silvestri, A., Vanhaecke, F., Muchez, P., Degryse, P., Isotopes on the beach Part 2 – Neodymium isotopic analysis for provenancing Roman glassmaking, 449-464, Copyright (2013), with permission from John Wiley and Sons.

The complete analytical procedures for the Sr-Nd isotopic analysis of sands are reported in De Muynck et al. (2009) and Ganio et al. (2012a, b). The results are shown in Appendix A, Fig. 3.1 and Fig. 3.2. The beach sands analysed show a wide range of $^{87}Sr/^{86}Sr$ isotope ratios between 0.7075 and 0.7290. The bulk concentration of Sr in the sands varies between 30 and 1186 ppm. ε_{Nd} values also vary significantly between -12.85 and -3.05. The Nd concentration of most sands analysed lies between 3 and 60 ppm Nd. One sample (IT21), however, contains 296 ppm Nd. The isotope ratio results and their relation to the composition of the beach sands and their broader geological setting are discussed in detail in Brems et al. (2013a,b).

3.5 $^{87}Sr/^{86}Sr$ as a provenance indicator of the lime source?

The large spread in $^{87}Sr/^{86}Sr$ isotope ratios encountered in this study can be attributed to two main sources of strontium. Calcareous fragments in the sand are the first important contributor. Shell fragments occur in widely varying amounts.

Sands with large proportions of shell material can contain high amounts of Sr. Some of the sands analysed are mainly composed of limestone grains. These show lower Sr contents. Shell and limestone fragments deliver Sr with low $^{87}Sr/^{86}Sr$ ratios to the sand. The $^{87}Sr/^{86}Sr$ isotope ratio of the shells is equal to that of present-day seawater, i.e., 0.709165 ± 0.000020 (Stille and Shields, 1997; Banner, 2004). Limestone has even lower $^{87}Sr/^{86}Sr$ isotope ratios with values depending on the age of the limestone and the extent of diagenesis (Burke et al., 1982). The broad range of $^{87}Sr/^{86}Sr$ isotope ratios, with Sr isotopic signatures mostly lying above the modern-day seawater value, implies that there must be a second Sr source bringing more radiogenic Sr to the system. The source of this radiogenic Sr must be sought in the silicate fraction of the sand and most likely in feldspar and, to a lesser extent, mica derived from crystalline magmatic or metamorphic rocks, or recycled immature sedimentary rocks.

The bulk Sr isotopic composition of the sand is a combined signal of the relatively unradiogenic Sr from the carbonates and the higher $^{87}Sr/^{86}Sr$ ratios from the aluminosilicates. This mixed Sr isotopic composition is not only dependent on the absolute content of carbonates and feldspar but also, and even to a much larger extent, on the proportion between the two. This relationship can be seen in Fig. 3.3. Whereas sands containing high CaO values generally have the lowest $^{87}Sr/^{86}Sr$ isotope ratios, these values can still be appreciably higher than the seawater value (Fig. 3.3a). The opposite can be seen in the $^{87}Sr/^{86}Sr$ vs. Al_2O_3 plot (Fig. 3.3b). Sands with low Al_2O_3 concentrations generally have $^{87}Sr/^{86}Sr$ isotope ratios very close to the modern-day seawater signature. As the Al_2O_3 concentration increases, the range of $^{87}Sr/^{86}Sr$ isotope ratios quickly widens. The $^{87}Sr/^{86}Sr$ vs. Al_2O_3/CaO plot (Fig. 3.3c) shows that $^{87}Sr/^{86}Sr$ isotope ratios of beach sands are only around or below modern-day seawater value, if they contain at least 4 times as much CaO as Al_2O_3 ($Al_2O_3/CaO < 0.25$) The higher the Al_2O_3/CaO ratio, the higher the $^{87}Sr/^{86}Sr$ isotope ratio.

In this regional study, only one sand sample was analysed for each beach deposit. Although care was taken to obtain a representative sample, small variations in the ratio of shell fragments to feldspar within the sand deposit can not be ruled out. These variations would result in small changes in the Sr isotopic signature of the sand.

Roman natron glass generally contains 5 to 10% CaO (Foster and Jackson, 2009). Taking into account the addition of natron and the loss of the volatile fraction during the melting of the glass, this corresponds to about 6 to 11% CaO in the sand raw material. Sands corresponding to this range of CaO concentrations have varying $^{87}Sr/^{86}Sr$ isotope ratios between 0.70851 and 0.71926 (Fig. 3.3a). However, not all of these sands are actually suitable to produce glass (see Chapter 2).

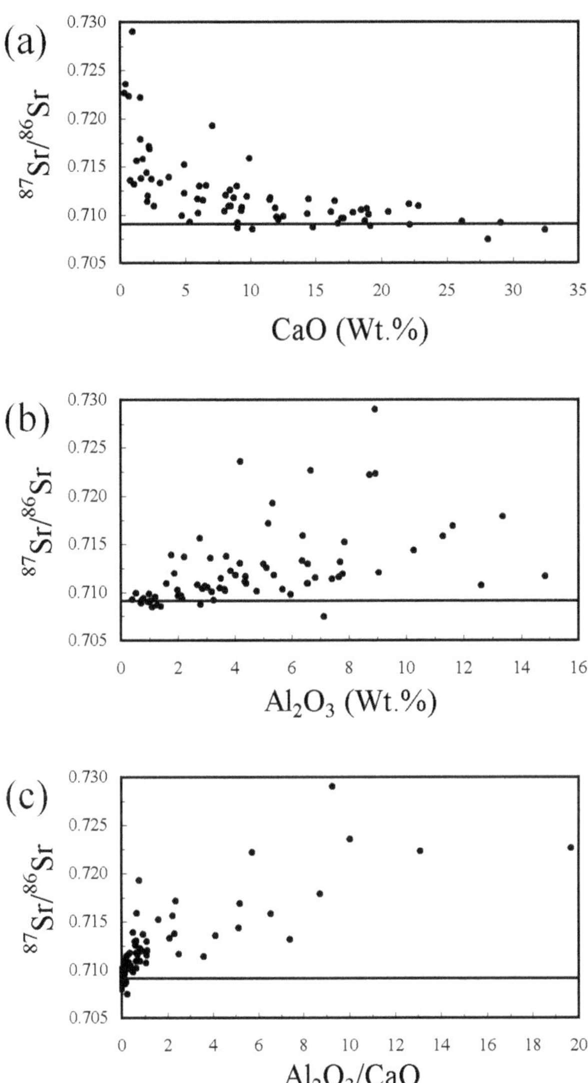

Figure 3.3:
(a) $^{87}Sr/^{86}Sr$ vs. CaO plot. (b) $^{87}Sr/^{86}Sr$ vs. Al_2O_3 plot. (c) $^{87}Sr/^{86}Sr$ vs. Al_2O_3/CaO plot. The thick black lines represent the present-day seawater $^{87}Sr/^{86}Sr$ isotope ratio of 0.709165 (Stille and Shields, 1997; Banner, 2004). Reprinted from Archaeometry, Vol 55, Brems, D., Ganio, M., Latruwe, K., Balcaen, L., Carremans, M., Gimeno, D., Silvestri, A., Vanhaecke, F., Muchez, P., Degryse, P., Isotopes on the beach Part 1 – Strontium isotope ratios as a provenance indicator for lime raw materials used in Roman glassmaking, 214-234, Copyright (2013), with permission from John Wiley and Sons.

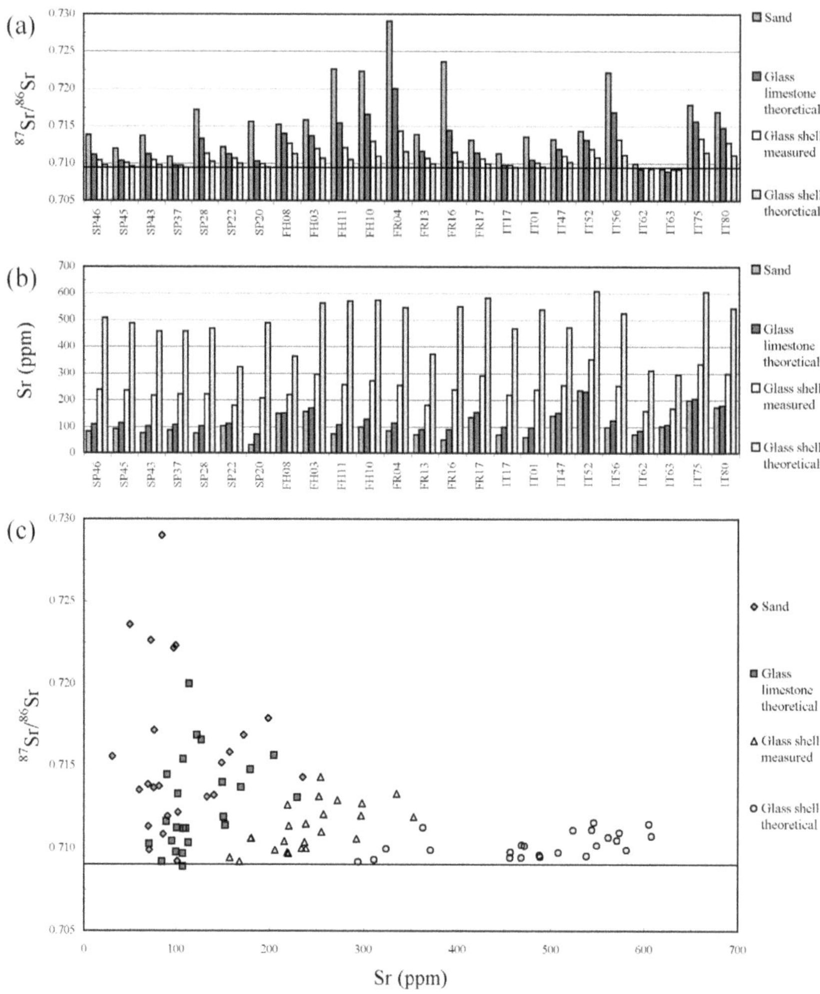

Figure 3.4:
(a) ^{87}Sr/^{86}Sr isotope ratio of sands with insufficient CaO to produce Roman natron glass and the glass that can be produced from these sands after the addition of an extra source of lime. (b) Sr concentrations. (c) ^{87}Sr/^{86}Sr vs. Sr concentration plot. The thick black lines represent the present-day seawater ^{87}Sr/^{86}Sr isotope ratio of 0.709165 (Stille and Shields, 1997; Banner, 2004). Reprinted from Archaeometry, Vol 55, Brems, D., Ganio, M., Latruwe, K., Balcaen, L., Carremans, M., Gimeno, D., Silvestri, A., Vanhaecke, F., Muchez, P., Degryse, P., Isotopes on the beach Part 1 – Strontium isotope ratios as a provenance indicator for lime raw materials used in Roman glassmaking, 214-234, Copyright (2013), with permission from John Wiley and Sons.

Most of them are too high in Al_2O_3 and Fe_2O_3 content, and do not contain enough SiO_2. Only one of the sands analysed would produce a typical Roman natron glass after fluxing it with pure natron (sand IT85 from the SE of Italy). Two other sands can be melted into glass resembling Roman glass for all but one element. Glass made with sand IT34 would have a low P_2O_5 concentration and glass melted with sand IT87 would be unusually low in Al_2O_3. The necessary amount of CaO in sands IT85 and IT87 is derived from limestones and marls in the local hinterland. The available CaO in sand IT34 is mostly contained in shell fragments naturally included in the sand. Therefore, we would expect the Sr isotopic signature of IT34 to resemble that of seawater, while those of IT85 and IT87 would have to be a little lower. However, the results are somewhat different. Sands IT85 and IT34 have Sr isotopic signatures slightly above the modern-day seawater value, i.e. 0.71079 and 0.71034, respectively. The $^{87}Sr/^{86}Sr$ isotope ratio of sand IT87 is indeed lower than the seawater signature: 0.70867. This is in accordance with its low Al_2O_3 content, which indicates that the contribution of radiogenic Sr from feldspar is very low. It therefore seems that (at least in the western Mediterranean, see below) the Sr isotopic signature is only indicative for the origin of the lime for sands (and glass) with low concentrations of Al_2O_3, or better low Al_2O_3/CaO ratios. As a limit value for this Al_2O_3/CaO ratio we can suggest 0.25 (e.g. for IT34 this value is 0.36, for IT85 0.29 and for IT87 0.14).

Twenty-four of the beach sands that were analysed for their Sr and Nd isotopic compositions contain insufficient calcareous fragments in order to provide sufficient CaO to produce a stable glass with a typical Roman composition (see Chapter 2). These sands with CaO concentrations below 6% mostly have rather radiogenic Sr isotopic signatures (Fig. 3.3a). The shortage of CaO in the sand could be compensated for by adding pieces of shell or limestone to the glass batch. By doing this, four of these lime-deficient sands (SP46, SP20, FR16 and IT01) could be used to produce glass with a composition very close to that of typical Roman natron glass. Depending on the Sr content and Sr isotopic signature of the material added, this could result in a shift in the $^{87}Sr/^{86}Sr$ isotope ratios of the final glass as compared to the Sr isotopic signature of the sand. In order to determine the extent of this possible shift, we calculated the expected Sr concentrations and Sr isotopic signatures of the hypothetical natron glasses. This was done by using binary mixing equations (Faure and Mensing, 2005). When two components with different Sr concentrations get mixed in varying proportions, the concentration of Sr in the resulting mixture is equal to:

$$Sr_M = Sr_A\, f_A + Sr_B\, (1 - f_A)$$

with Sr_M, Sr_A and Sr_B representing the Sr concentrations in the mixture, component A and component B, respectively, and f_A and $(1 - f_A)$ expressing the fractions of component A and B, respectively. If these two components also have different isotopic signatures, the $^{87}Sr/^{86}Sr$ isotope ratio of the final mixture can be calculated with the following equation (Faure and Mensing, 2005):

$$\left[\frac{^{87}Sr}{^{86}Sr}\right]_M = \frac{\left[\frac{^{87}Sr}{^{86}Sr}\right]_A f_A \; Sr_A + \left[\frac{^{87}Sr}{^{86}Sr}\right]_B (1-f_A) \; Sr_B}{Sr_A \, f_A + Sr_B \, (1-f_A)}$$

For each lime-deficient sand, three calculations were performed, each one simulating glass production using a different source of additional lime. In the first calculation, the additional lime needed was assumed to come from shell material with a high Sr content of 4000 ppm and a $^{87}Sr/^{86}Sr$ isotope ratio equal to that of modern-day seawater: 0.709165 (Stille and Shields, 1997; Banner, 2004). For the second calculation, we used the Sr concentration and $^{87}Sr/^{86}Sr$ isotope ratio that was actually measured for shells collected from a number of beaches along the coasts of southern France and northwest Italy as we would expect shells to have been collected by a local Roman glass producer. The shells were crushed, homogenised and analysed for their Sr concentration and $^{87}Sr/^{86}Sr$ isotope ratio in the same way as the beach sands. This resulted in a lower Sr concentration of 1568 ppm. The $^{87}Sr/^{86}Sr$ isotope ratio remains practically the same: 0.70915. For the third calculation, we used theoretical limestone with a Sr content of 400 ppm and $^{87}Sr/^{86}Sr$ ratio of 0.70750. The composition of the hypothetical glasses and the proportion of sand to additional lime was calculated as described in Chapter 2: Na_2O and CaO levels were fixed at 16.63 and 7.48%, respectively (the average Na_2O and CaO content of Roman glass; Foster and Jackson, 2009). We further assumed that the contribution of natron to the Sr budget of the glass is negligible.

The original Sr concentrations and $^{87}Sr/^{86}Sr$ isotope ratios of the lime-deficient sands and the calculated Sr contents and $^{87}Sr/^{86}Sr$ isotope ratios of the glasses are shown in Fig. 3.4. Most of the sands with low CaO concentrations initially have rather high and variable $^{87}Sr/^{86}Sr$ ratios (0.70922 – 0.72901) and relatively low Sr contents (30 – 236 ppm). After the addition of shell or limestone, the Sr isotopic signature of the resulting glass is shifted towards lower $^{87}Sr/^{86}Sr$ ratios (Fig. 3.4a). The extent of this shift is strongly dependent on the concentration of Sr in the source of additional lime. Shell with a high Sr content of 4000 ppm would have the largest influence. Glasses made with this kind of lime source would have $^{87}Sr/^{86}Sr$ isotope ratios between 0.70918 and 0.71158. For glasses made with the

French and Italian seashells with a Sr concentration of 1568 ppm, the $^{87}Sr/^{86}Sr$ isotope ratio is less modified (0.70919 – 0.71433). The calculated $^{87}Sr/^{86}Sr$ isotope ratios of glasses produced with limestone are on average even higher and show a wider range (0.70889 – 0.71999), even though we used a low $^{87}Sr/^{86}Sr$ isotope ratio of 0.70750 for the limestone in the calculations. Only for one sample (IT63) the final $^{87}Sr/^{86}Sr$ isotope ratio is lower than the modern-day seawater value. This sand already had a relatively low $^{87}Sr/^{86}Sr$ isotope ratio to start with. It therefore seems that the Sr isotopic signature is not necessarily indicative for the source of lime in a glass. Our binary mixing calculations demonstrate that glass produced with Sr-rich seashell can still show a significant difference between its $^{87}Sr/^{86}Sr$ isotope ratio and that of seawater of up to 0.00241. Furthermore, we see that glass produced with these sands and limestone usually has higher $^{87}Sr/^{86}Sr$ ratios than those made with shell fragments.

The concentration of Sr in the calculated glasses is shown in Fig. 3.4b and 3.4c. It seems that the Sr concentration is more diagnostic for the source of lime. Glasses produced with shells with Sr contents of 4000 ppm would contain between 294 and 608 ppm Sr. Glasses made with limestone with 400 ppm Sr have Sr concentrations between 70 and 230 ppm. The range of Sr concentrations in glasses made with shell material with intermediate Sr contents, however, overlaps with both previous ranges. The use of the Sr content of glass as a provenance indicator of the lime source used thus seems to be limited by the wide possible range of Sr content in (partially recrystallised) aragonite and high-Mg calcite, which make up the shells on modern-day beaches.

Freestone et al. (nd) previously determined the Sr isotopic composition of raw glass from primary glass production sites in Egypt and Syro-Palestine, which were active between the 4th and 8th century AD. The Sr isotopic signature of the Levantine raw glass was found to be very close or slightly below the modern-day marine signature. Furthermore, they have high Sr contents between 300 and 500 ppm, suggesting the use of shell as source of lime. The Egyptian samples, however, had lower $^{87}Sr/^{86}Sr$ ratios and Sr contents between 100 and 200 ppm. These two observations are in favour of a limestone source of lime (Freestone et al., 2003). The same combination of seawater Sr isotopic signatures and high Sr concentrations, and low $^{87}Sr/^{86}Sr$ isotope ratios and low Sr contents has been observed in numerous analyses of Roman natron glass (e.g., Wedepohl and Baumann, 2000; Freestone et al., 2003; Degryse et al., 2006a, b, c; 2008, 2009b; Degryse and Schneider, 2008). Then why doesn't this approach seem to work for our calculated glasses? The problem seems to be that the Sr isotopic signature of the lime source is overshadowed by the influence of the Sr from the silicate fraction in the sands. In the western Mediterranean, Sr derived from the silicates is usually very radiogenic. Lithogenic (i.e. non-carbonate) Nile sediments, which dominate the eastern Mediterranean, however, have lower $^{87}Sr/^{86}Sr$ ratios that are very close to or lower than the present-day seawater value, with a pure Nile end-member of 0.707043 and sediments along the Nile delta and the Levantine coast varying between 0.7075 and 0.7095 (Krom et al., 1999a, b; Weldeab et al., 2002). In the eastern Mediterranean, the influence of Sr from feldspar would thus be much smaller and the final Sr isotopic signature of the glass would be left unaltered or only slightly lowered. Therefore, we conclude that Sr isotopic signatures are indeed indicative of the source of lime for natron glass produced in the eastern Mediterranean, whereas for glass made in the west this is not (always) the case. Sr concentrations provide the same information in both regions.

3.6   The ε_{Nd} value as a provenance indicator for the silica source?

Nd concentrations of the analysed sands show a (weak) positive correlation with Fe_2O_3, TiO_2 and P_2O_5 contents. This suggests that Nd indeed comes into the glass with the non-quartz (heavy) mineral fraction of the sand, but that the ε_{Nd} value is a mixture of all these mineral species. ε_{Nd} values are not correlated with any of the major or minor elements. The ε_{Nd} values of the beach sand analysed are broadly related to the large geological regions and vary relatively gradually along the coastlines (Fig. 3.2). Spanish and French sands all show relatively low ε_{Nd} values from -12.4 to -8.0 in close agreement with the data from the deep sea sediments. Italian sands, however, show a wide range of ε_{Nd} values between -12.8 and -3.0.

The silicate fractions of most of the analysed sands are the disintegration products of relatively old Hercynian metamorphic and magmatic rocks, or younger sedimentary rocks, which ultimately have the same origin. The ε_{Nd} values of these sediments are consequently rather low, mostly between -12.5 and -8.0. Sands occurring close to the oldest magmatic and metamorphic complexes show the most radiogenic Nd isotopic signatures. This is the case along the Betics Cordillera, the Maures-Tanneron massif, Calabria and NE Sicily. The sands with the most negative ε_{Nd} values below -12.5, however, were found on the northwesternmost part of Sicily (IT62, IT63) and are derived from Miocene sedimentary rocks. The sediments deposited in this area during the Miocene were probably derived from old (Paleozoic or even Precambrian) rocks with a crustal affinity.

Along the Mediterranean coastline of Spain, France and Italy, we can identify three regions where ε_{Nd} signatures of beach sands are relatively elevated to values above -7: in the Gulf of Genoa, in the area near the Vesuvius (Naples) and the Apulia Region. High ε_{Nd} values (up to -3.05) in the Gulf of Genoa are attributed to the occurrence of Jurassic metaophiolites in that region. The ultramafic source of sands in this area results in very high Fe_2O_3, MgO and Al_2O_3 concentrations. As a consequence, these sands cannot have been used to produce Roman natron glass.

The occurrence of a broad range of Nd isotopic signatures in the area near the Vesuvius and the Campanian coastal strip can be accounted for by local recent volcanic activity. These sands are derived from Pleistocene–Holocene rhyolitic volcanic rocks and contain high percentages of heavy minerals, resulting in high Fe_2O_3 and Al_2O_3 levels, making them unsuitable as raw materials for glass production (see Chapter 2). ε_{Nd} values of sands from the Campanian beaches, which get delivered to the coast by the Volturno River, can vary greatly because of the highly variable Nd isotopic signatures of the Roccamonfina and Somme-Vesuvius volcanic rocks (ε_{Nd} from -11 to 0; Conticelli et al., 2002; and references therein). A similar rise in ε_{Nd} values is observed near the Etna on Sicily. The sand of this area that was analysed (IT71), however, only contained a very small amount of Etna-derived grains and consequently its ε_{Nd} value is only moderately elevated to -7.86. Sandy sediments closer to the Etna volcano, will most likely show higher ε_{Nd} values.

The ε_{Nd} values between -6.11 and -4.17 that occur in the Apulia Region in SE Italy can be attributed to the presence of mafic minerals (mostly pyroxenes) either directly or indirectly derived from Mount Vulture volcanic rocks (ε_{Nd} values between 1 and 3; Conticelli et al., 2002; and references therein). The area near the Vulture volcano gets drained by the Ofanto River which discharges into the Manfredonia Gulf. This results in relatively unradiogenic Nd isotopic signatures for sands in the gulf (IT90). The Ofanto River sediments get further transported to the south along the Apulian coast at least up to the Cape of Otranto (Caldara et al., 1998; Mastronuzzi et al., 2007). The heavy mineral fraction of these sediments contributes to the beach sands along the coast which consequently show relatively high ε_{Nd} values (IT87). The Pleistocene sedimentary rocks that overlay the platform carbonates of the Adria microplate contain a number of layers rich in mafic minerals which originate from Mount Vulture (Acquafredda et al., 1997; Mastronuzzi and Sansò, 2002). Recycling of these sedimentary sequences also delivers unradiogenic Nd to the beaches in the area (IT85, IT87).

When we focus on the sands assessed suitable for Roman natron glass production (see Chapter 2) we can evaluate the Nd isotopic signature as a provenance indicator for sand raw materials and raw natron glass. The four sand raw materials that can be used to produce Roman natron glass after the addition of extra lime all have relatively low ε_{Nd} values, typical for the western Mediterranean. Sand SP46, from southwestern Spain, has an ε_{Nd} value of -7.99. Suitable sand from near El Rubial in southeastern Spain (SP20) has a ε_{Nd} value of -11.76. Sand raw materials from Les Bormettes in the Provence (FR16) have an even lower ε_{Nd} value of -12.40. Finally, the sands from Calla Violina in Tuscany, which would make good natron glass after the addition of lime, have ε_{Nd} values of -8.86. The mixture of shell fragments, which were collected along several beaches in S France and NW Italy, has a Nd concentration of 1.0 ppm and ε_{Nd} value of -6.6. Since this Nd concentration is 5 to 20 times smaller than the concentration in most sands and the additional lime only makes up ~10% of the total glass batch, the additional source of lime only has a very small influence on the final Nd budget of a glass. Glass produced with either of these sand sources thus would be readily distinguishable from glass from the eastern Mediterranean since raw glass from Egypt and Syro-Palestine has relatively low variation in ε_{Nd}, with values between -6 and -4 (Degryse and Schneider, 2008; Freestone et al., nd; this study).

Three sands from Italy were identified as being suitable for Roman glass production in their present form. Sand IT34 from Tuscany also has a rather low ε_{Nd} value of -9.42. The two other suitable glassmaking sands found in the Basilicata (IT85) and Apulia Regions (IT87) in SE Italy, however, have relatively high ε_{Nd} values of -6.1 and -4.2, respectively. These values coincide with the range of Nd isotopic signatures previously thought to be characteristic for an eastern Mediterranean origin (Degryse and Schneider, 2008; Freestone et al., nd). This shows that the use of Nd isotopic signatures as a provenance indicator for Roman glass is not as straightforward as previously thought. However, based on the Nd isotopic signature of glass artefacts, certain possible areas of production can be suggested or excluded. For a further differentiation, other techniques, such as trace element analysis, will have to be applied.

3.7   Conclusions

The aim of this chapter was to evaluate the use of Sr and Nd isotopic signatures for the provenance determination of Roman natron glass. To do this, 77 beach sands from the coasts of Spain, France and Italy were analysed for their Sr and Nd isotopic compositions. By using binary mixing equations, we were able to calculate the Sr isotopic signature of glass produced from the sands analysed. The results of these calculations show that the addition of shell or limestone to the glass batch is often not enough to obscure the radiogenic Sr isotopic signature of the sand raw materials. Therefore, the $^{87}Sr/^{86}Sr$ isotope ratio of glass is not always indicative of the main source of lime. In this aspect, there seems to be a marked difference between the eastern and western Mediterranean. In the west, the silicate fraction of the sand often contains important amounts of radiogenic Sr, resulting in a shift in $^{87}Sr/^{86}Sr$ ratios to higher values. Since the Sr isotopic signatures of the Nile-dominated siliciclastic sediments in the eastern Mediterranean lie around or slightly below the modern-day seawater signature, the induced shift in the Sr isotopic signature of the glass is less explicit. For glasses produced in the eastern Mediterranean, the Sr isotopic signature is indeed a good provenance indicator for the source of lime. In the western part of the Mediterranean, this is only so for glasses with low Al_2O_3 concentrations and an Al_2O_3/CaO ratio lower than 0.25. The Sr content of the glass appears to be a better and more robust indication of the lime source.

Nd in Roman natron glass originates from the non-quartz (heavy) mineral fraction of the sand raw material and its isotopic composition is an indication for the source of the silica. Suitable glassmaking sands from Spain, France and the western part of Italy all have relatively low ε_{Nd} values and glass that would be produced from them could be readily distinguished from glass from the known primary production sites in Egypt and Syro-Palestine. Two good sand sources in the Basilicata and Apulia Regions (SE Italy), however, have ε_{Nd} values that may coincide with those of glass with an eastern Mediterranean origin. In Chapter 4, we will determine whether these possible sources of glass can be distinguished from each other using trace element patterns.

Trace elements in sand raw materials

D. Brems, P. Degryse

As discussed in previous chapters, the major elemental composition of Roman natron glass is relatively uniform and not very useful for provenance determination. The use of radiogenic isotopes of Sr and Nd was examined in Chapter 3. There we have seen that isotopic compositions can be very useful indicators but are usually insufficient to pinpoint the origin of a glass artefact. It was shown that certain sands from Italy, which were found to be suitable for Roman natron glass production, have ε_{Nd} values that coincide with those of sand raw materials and the glass from known primary production centres in the eastern Mediterranean. In this chapter, we will determine whether a distinction between the different possible source areas can be made using trace element geochemistry.

4.1 Trace elements in Roman glass

Basic Roman natron glass can be seen as a mixture of three components: silica sand, lime-bearing material and natron as the soda-rich flux. Additionally, glass was often coloured or decoloured by adding a small amount of specific minerals. These raw materials all introduced a number of trace elements to the glass batch (Fig. 4.1). In particular those solely related to the sand raw material are of interest as possible provenance indicators.

4.1.1 Natron

Natron was a relatively pure source of soda. Depending on the mineralogy of the evaporitic deposits, the concentration of elements such as Cl and S varies greatly (Brill, 1999; Currie, 2008; Shortland, unpublished data). These elements were probably almost entirely introduced to the glass batch by the natron flux. The concentration of Cl and S in the final glass is, however, limited by their solubility in soda-lime-silica melts (Bateson and Turner, 1939; Gerth et al., 1998; Köpsel, 2001; Shugar and Rehren, 2002; Salviulo et al., 2004). The concentration of most other commonly measured trace elements in both modern and ancient evaporites from the Wadi Natrun is found to be very low (Currie, 2008; Wedepohl et al., 2011b; Shortland, unpublished data). When compared with the concentration of trace elements in natron glass as given by, for example, Degryse and Shortland (2009) and Wedepohl et al. (2011a), only B, P, Br and U appear to occur in the same order of magnitude in both natron and natron glass. Mg concentrations are 30 times lower in the flux. The concentration of elements such as Li, K, V, Cr, Ni, Cu, Zn, Sr, Zr, Ba, Pb and the rare earth elements (REE) are between 60 and 1500 times lower in natron than in average natron glass (Currie, 2008; Wedepohl et al., 2011b; Shortland, unpublished data).

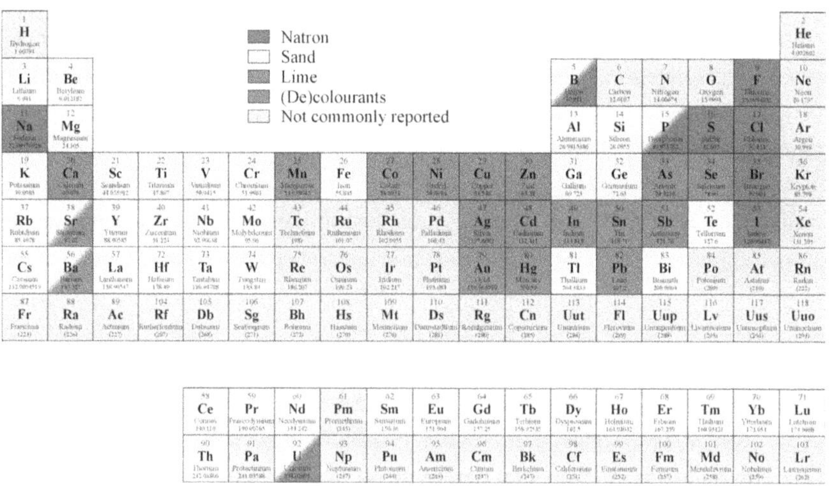

Figure 4.1:
Periodic table with indication of the most likely sources of the elements in Roman natron glass. Reprinted from Archaeometry, Vol 56, Brems, D., Degryse, P., Trace element analysis in provenancing Roman glass-making, 116-136, Copyright (2014), with permission from John Wiley and Sons.

4.1.2 Sand

Because of the small ionic size of Si^{4+} (0.026 nm; Shannon, 1976), only very small amounts of other elements such as Al, Ti, Fe and Ge can be incorporated into the crystal structure of quartz. Minor and trace elements in quartz-rich sands are generally concentrated in associated accessory minerals such as feldspar, pyroxene, amphibole, zircon, Fe-Ti oxides, monazite and clay minerals, among others.

Sand deposits can contain different types and quantities of accessory minerals, depending on the nature of the source rocks and the local geological setting. Different mineralogical compositions of the sands will lead to different trace element signatures, which may be useful tracers for their origin. The application of trace elements as provenance indicators for ancient glass was explored in some recent studies. Roman natron glass generally contains relatively low levels of trace elements. This is attributed to the use of mineralogically mature sand, rich in quartz and relatively depleted in heavy minerals (Freestone et al., 2000, 2002a). HIMT glass typically contains higher concentrations of trace elements, suggesting the use of less pure quartz-sands (Freestone et al., 2005). Promising groupings of analysed glass fragments could be made based on elements like Zr-Ti (Aerts et al., 2003), Zr-Ti-Cr-La (Shortland et al., 2007; Reade, 2009; Walton et al., 2009) and Zr-Sr-Ba (Freestone et al., 2000; Paynter, 2006; Silvestri, 2008; Silvestri et al., 2008). These elements are usually not related to any colouring agents that could have been deliberately added to the glass batch and can be used to distinguish between different sand raw materials. Zr, together with Hf, is accumulated in the heavy mineral zircon. Trace elements such as Sc, Ti, V and Cr are generally correlated with iron oxides (Wedepohl et al., 2011a) or with specific heavy minerals like rutile, ilmenite, titanite and chromite. Ba and Rb are related to alkali feldspar. Wedepohl and Baumann (2000) attributed relatively high Ba concentrations in Roman glass from Hambach to the presence of barite as a heavy mineral in the glassmaking sand. Ga substitutes for Al in aluminium silicates (Wedepohl et al., 2011a). Other elements which are probably also related to the sand raw materials are Li, Be, Ge, Y, Nb, Mo, Te, Cs, La, Ta, W, Tl, Bi and Th (Shortland et al., 2007; Degryse and Shortland, 2009; Reade, 2009; Wedepohl et al., 2011a, b). B, P and U are partly derived from the silica source (Shortland et al., 2007), but they also come in with the natron flux. Sr is provided by both the sand (i.e., mostly feldspar and mica) and the source of lime (Brems et al., 2013a). In Mn-rich glasses, some of the Sr may be introduced through the Mn ore (Ganio et al., 2012b, c).

Rare earth elements (REE) are classically used as provenance indicators of sediments and sedimentary rocks (e.g. McLennan, 1989; Lidiak and Jolly, 1996). In quartz-rich sands, these elements are mostly concentrated in the clay and silt fraction (Cullers et al., 1979; Tlig and Steinberg, 1982; McLennan, 1989; Yokoo et al., 2004). However, heavy mineral species can also contain significant amounts of REE. The light REE are accumulated in monazite and allanite, while the heavy REE are relatively concentrated in zircon and garnet (Gromet and Silver, 1983; McLennan, 1989; McKay, 1989). Unlike the other REE, Eu can occur in a divalent state and as a result it can be preferentially incorporated in plagioclase. Enrichment or depletion in plagioclase during weathering, erosion or sedimentary processes, can cause positive or negative Eu-anomalies in the REE patterns of sandy sediments, which in turn can be passed on to the glass (for a more extensive discussion see also Degryse and Schneider, 2009 and Wedepohl et al., 2011b). Average REE patterns appear to be distinctly different between three major ancient glass groups, i.e. soda ash glass, natron glass and wood ash glass (Wedepohl et al., 2011b). However, within the group of natron glass, REE patterns are relatively uniform (Degryse and Shortland, 2009; Wedepohl et al., 2011b). This would indicate that the REE are derived from the clay fraction of the sand raw materials or from interaction with the furnace walls, and are of little use as a provenance indicator for Roman natron glass (Degryse and Shortland, 2009; Walton et al., 2009).

4.1.3   Lime

Sr in natron glass is mostly derived from the shell or limestone introduced (whether or not deliberately) as the source of lime (Wedepohl and Baumann, 2000; Freestone et al., 2003; Brems et al., 2013a). High Sr concentrations in glass suggest the use of shell fragments, while low Sr contents indicate the use of limestone. However, other mineral species in the sand raw material, such as feldspar and mica, can also introduce Sr to the glass. The lime-bearing component of the glass batch can also introduce minor amounts of chemically related elements such as Mg, Fe and Mn. Rare earth elements and other trace elements only occur in very small concentrations in shell material (e.g. Wedepohl et al., 2011a, b).

4.1.4   (De)colourants and recycling

Pure soda-lime-silica glass, without any impurities, is essentially colourless. Most ancient glass fragments, however, contain a small amount of Fe_2O_3. This results in a green or blue (also: aqua) tint. This iron was usually unintentionally introduced as impurities in the sand raw material. The green-blue colour could be neutralised by the addition of Mn or Sb, which resulted in the oxidation of the Fe^{2+} to the practically colourless Fe^{3+} (Sayre, 1963; Sayre and Smith, 1961; Brill, 1988). By adding different metals in varying concentrations and under different furnace conditions, a whole range of different colours could be achieved. Details on the effects of these different elements on the colour of glass are beyond the scope of this study and can be found in Weyl (1951), Bamford (1977), Green and Hart (1987), and Pollard and Heron (2008).

Elements commonly associated with (de)colouring in ancient glass are Mn, Co, Ni, Cu, Zn, As, Se, Ag, Cd, In, Sn, Sb, Au, Hg and Pb (Aerts et al., 2003; Shortland et al., 2007; Degryse and Shortland, 2009; Reade, 2009; Wedepohl et al., 2011a). Some of these elements do not influence the colour of the glass, but they occur as impurities in the mineral (de)colourants. Elevated concentrations of these elements (over 1000 ppm) suggest they were deliberately added to the glass batch to influence the colour of the resulting product. Concentrations between about 100 ppm and 1000 ppm are typically interpreted as indications for glass recycling (Freestone et al., 2002a; Silvestri et al., 2005; Degryse and Shortland, 2009; Foster and Jackson, 2010). Remelting a batch of colourless cullet with small amounts of coloured fragments would result in concentrations of these colouring elements not high enough to significantly alter the colour and suggest intentional addition, but too high to be explained by natural impurities in the sand raw materials. Low concentrations (in the 1 – 100 ppm range) of trace elements like Co, Cu, and Pb suggest that the glass was produced from primary raw materials and that recycling was limited, or that recycling took place after very careful selection of cullet to avoid contamination (Silvestri, 2008; Silvestri et al., 2008). These concentration boundaries are however arbitrary. Little is known about the actual background concentrations of (de)colourant-related elements in glassmaking sand raw materials.

Roman natron glasses decoloured by Mn often show elevated Ba contents and a strong positive correlation between Mn and Ba (Brill, 1988; Paynter, 2006; Foster and Jackson, 2010). This is consistent with the use of wad or a mixture of pyrolusite (MnO_2) and psilomelane ($(Ba,H_2O)_2Mn_5O_{10}$) as the source of Mn (Jackson, 2005; Silvestri, 2008). Next to Ba, the Mn source may also introduce extra Sr to the glass (Ganio et al., 2012b, c).

4.2   Trace elements in sand raw materials

The trace elemental composition of natron glass will be a combination of elements derived from the sand, the natron and the lime, and possibly from any (de) colourants that were added to the glass batch. In this chapter, we investigate the variation in trace element signatures of suitable sand raw materials. We examine whether the elements commonly related to the sand raw material can be used to distinguish between sand sources with similar Nd isotopic signatures. The concentrations of trace elements generally associated with (de)colouring agents will provide more information about the background levels for these elements that are attributable to the sand raw materials. 11 beach sand samples were analysed for their trace elemental compositions. These sands were found to be the most suitable for Roman natron glass production (see Chapter 2). A mixture of shell material collected from several beaches along the Mediterranean coast of France and NW Italy was also included in the analysis to investigate the possible influence of the

	SiO$_2$	Al$_2$O$_3$	Fe$_2$O$_3$	MgO	CaO	Na$_2$O	K$_2$O	P$_2$O$_5$	Cl	TiO$_2$	MnO	Sc	V	Cr	Co
SiO$_2$	1,00														
Al$_2$O$_3$	0,03	1,00													
Fe$_2$O$_3$	-0,12	0,58	1,00												
MgO	-0,50	0,26	0,72	1,00											
CaO	-0,90	-0,43	-0,21	0,26	1,00										
Na$_2$O	-0,07	0,94	0,36	0,05	-0,29	1,00									
K$_2$O	-0,10	0,05	-0,25	-0,42	0,11	0,16	1,00								
P$_2$O$_5$	-0,11	0,04	0,57	0,54	0,00	-0,08	-0,50	1,00							
Cl	0,31	0,14	0,12	-0,23	-0,40	0,22	0,02	0,20	1,00						
TiO$_2$	0,49	0,35	0,33	-0,15	-0,58	0,29	-0,11	0,20	0,10	1,00					
MnO	-0,76	-0,04	0,24	0,33	0,71	-0,01	0,39	0,12	-0,45	-0,20	1,00				
Sc	-0,26	0,46	0,91	0,74	-0,01	0,23	-0,03	0,56	-0,13	0,23	0,49	1,00			
V	0,34	0,42	0,68	0,35	-0,50	0,19	-0,25	0,46	-0,07	0,77	-0,10	0,66	1,00		
Cr	0,09	-0,07	-0,18	-0,19	0,01	-0,11	0,27	-0,70	-0,52	-0,03	0,17	-0,09	-0,12	1,00	
Co	-0,02	0,42	0,90	0,56	-0,23	0,20	-0,06	0,66	0,17	0,44	0,27	0,90	0,76	-0,33	1,00
Zn	0,31	0,20	0,42	-0,14	-0,42	0,19	-0,05	0,16	0,77	0,42	-0,25	0,13	0,26	-0,26	0,44
Br	0,31	0,75	0,42	-0,58	-0,49	0,99	0,66	-0,30	0,98	0,89	-0,13	0,26	0,26	-0,30	0,55
Rb	0,02	-0,21	-0,30	-0,36	0,06	-0,14	0,85	-0,34	0,20	-0,30	0,24	-0,13	-0,41	0,12	-0,08
Sr	-0,80	-0,32	-0,39	0,05	0,89	-0,11	0,12	-0,17	-0,31	-0,52	0,45	-0,25	-0,52	-0,08	-0,42
Zr	-0,28	0,07	0,44	0,29	0,21	0,07	-0,02	0,64	-0,33	0,59	0,71	0,52	0,46	0,70	0,48
Sb	0,19	0,17	0,38	-0,12	-0,32	0,20	0,03	0,07	0,80	0,19	-0,23	0,12	0,05	-0,16	0,34
Cs	-0,03	-0,14	0,05	0,16	0,07	-0,24	-0,15	0,01	-0,24	-0,45	0,16	0,03	-0,29	0,10	-0,02
Ba	-0,67	-0,05	-0,15	0,15	0,64	0,11	0,18	-0,03	-0,02	-0,36	0,29	-0,01	-0,24	-0,20	-0,17
Nd	-0,36	-0,40	0,15	0,16	0,47	-0,39	0,00	0,44	-0,06	0,12	0,43	0,33	0,26	-0,20	0,33
Eu	-0,68	0,11	0,24	0,14	0,58	0,24	0,21	0,18	0,05	0,02	0,59	0,31	0,03	-0,26	0,28
Tb	-0,60	-0,10	0,41	0,31	0,56	-0,10	0,13	0,41	-0,11	0,08	0,73	0,54	0,24	-0,21	0,53
Dy	-0,53	-0,02	0,59	0,52	0,45	-0,11	-0,06	0,54	-0,17	0,11	0,66	0,73	0,42	-0,14	0,64
Yb	-0,51	0,26	0,72	0,68	0,30	0,17	-0,16	0,69	-0,08	0,22	0,56	0,82	0,51	-0,26	0,71
Lu	-0,42	0,28	0,76	0,66	0,20	0,16	-0,13	0,66	-0,05	0,28	0,51	0,86	0,59	-0,21	0,76
Hf	-0,06	0,02	0,27	0,01	0,05	0,04	-0,08	0,48	-0,06	0,63	0,37	0,34	0,43	-0,05	0,37
Ta	0,17	0,44	0,52	0,05	-0,34	0,40	-0,13	0,30	0,15	0,80	0,04	0,45	0,68	0,05	0,50
Th	-0,24	-0,15	0,34	0,18	0,25	-0,15	0,01	0,42	0,02	0,30	0,49	0,45	0,27	0,09	0,38

Table 4.1:
Correlation matrix for major, minor and trace elements analysed by INAA and ICP-OES.

Zn	Br	Rb	Sr	Zr	Sb	Cs	Ba	Nd	Eu	Tb	Dy	Yb	Lu	Hf	Ta	Th
1,00																
0,87	1,00															
0,00	0,55	1,00														
-0,40	-0,37	-0,05	1,00													
0,08	0,55	-0,22	-0,04	1,00												
0,96	0,89	0,09	-0,30	0,07	1,00											
-0,16	-0,48	0,16	-0,16	-0,24	-0,16	1,00										
-0,23	-0,13	-0,08	0,82	-0,11	-0,09	-0,55	1,00									
0,09	-0,01	-0,14	0,39	0,51	0,12	-0,45	0,52	1,00								
0,22	0,32	-0,10	0,63	0,43	0,33	-0,38	0,69	0,67	1,00							
0,21	0,26	-0,05	0,43	0,62	0,26	-0,16	0,42	0,82	0,86	1,00						
0,14	0,07	-0,23	0,27	0,64	0,21	-0,19	0,39	0,84	0,73	0,92	1,00					
0,09	0,18	-0,33	0,12	0,78	0,15	-0,29	0,35	0,66	0,61	0,74	0,89	1,00				
0,14	0,30	-0,31	0,03	0,75	0,20	-0,33	0,31	0,66	0,57	0,72	0,89	0,99	1,00			
0,14	0,69	-0,25	-0,09	0,99	0,10	-0,41	-0,02	0,57	0,36	0,49	0,55	0,63	0,62	1,00		
0,44	0,76	-0,42	-0,34	0,82	0,40	-0,59	-0,04	0,36	0,32	0,31	0,44	0,58	0,63	0,79	1,00	
0,20	0,79	-0,10	0,02	0,84	0,50	-0,40	0,19	0,74	0,45	0,61	0,73	0,73	0,74	0,84	0,67	1,00

addition of extra shell fragments on the final bulk trace element signature of a glass batch.

Sand and shell samples were finely crushed in an agate mortar and analysed for their trace elemental composition using Instrumental Neutron Activation Analysis (INAA). Complete analytical procedures are reported in Brems and Degryse (2014).

Results of the trace element analysis of the suitable glassmaking sands and the shell fragments are listed in Appendix A. Table 4.1 shows the correlation matrix for the trace elements analysed by INAA and major and minor elements previously determined via ICP-OES. Elements, for which the concentration was above the detection limit for only three samples or less, were discarded. For Br, measurable concentrations were detected in only four sand samples, so any correlations found with this element must be evaluated carefully to rule out any errors due to the small amount of data points.

4.2.1   Natron-related elements

The elements Na, Cl, S, P, F, Br, I, B and U in natron glass are fully or partially attributed to the natron flux used (Fig. 4.1). Of these elements, Na and P were previously analysed via ICP-OES and discussed in Chapter 2. Cl, Br and I could be analysed with INAA. Cl concentrations in the analysed sand samples vary between 0.01 and 0.53%. This variation is a result of the varying amounts of halite from the seawater adhered to the beach sand grains. Concentrations of Br in the beach sands range from below detection limit to 20 ppm and are strongly correlated with Cl (correlation coefficient r = 0.984). I was found to be below the detection limit for all samples analysed except for sand SP45, which contains 1.1 ppm I. Although the shell fragments were washed to remove sand and salt particles before crushing, they still contain 0.12% Cl indicating its presence as inclusions in the shell itself. These results show that beach sand raw materials and seashell fragments can introduce some 'natron-related' elements such as Cl and Br to the glass batch.

4.2.2   Sand-related elements

The elements that are exclusively related to the sand raw materials are of particular interest for provenance studies of ancient glass. Two of those elements are Zr and Hf as they are almost entirely derived from the heavy mineral zircon. The Zr contents of the sands and shell analysed range from below the detection limit to 198 ppm. It must be noted, however, that the reported detection limits are rather high: < 16 ppm for the shell material and < 68 to < 220 ppm for the sand samples. Concentrations of Hf vary between 1.01 and 5.35 ppm in the sand. Shell material only contains 0.03 ppm Hf. Zr and Hf are very strongly correlated (r = 0.995) with Zr/Hf ratios between 30 and 40, typical for zircons derived from granites (Gulson, 1969).

Next to Zr and Hf, TiO_2 is also generally related to the heavy mineral fraction in the sand raw materials and as a result they too are often correlated. TiO_2 contents range from 0.08 to 0.61%. In a biplot of TiO_2 vs. Zr, the expected correlation between the two elements is indeed present (r = 0.948; Fig. 4.2a). Sand sample IT85, however, does not follow the same correlation and is relatively enriched in Zr, reflecting a relative enrichment of zircon with respect to the Ti-rich mineral species, such as rutile, ilmenite and/or titanite, in the sand. Sc concentrations in the sands vary between 1.17 and 4.30 ppm. V contents lie between 12.8 and 32.9 ppm. Moderate to good correlations of these elements with Fe_2O_3 and TiO_2 suggest they are also related to the heavy mineral fraction of the sand. The concentration of Ta in the beach sands varies between 0.18 and 0.48 ppm. Ta is relatively well correlated with other elements of group 4 and 5 of the periodic table such as TiO_2 (r = 0.801), V (r = 0.676), Zr (r = 0.825) and Hf (r = 0.786). The most common heavy mineral containing Cr, is chromite. This mineral is generally associated with ultramafic igneous rocks. Cr concentrations in the sands analysed in this study vary widely between 8.2 to 277 ppm.

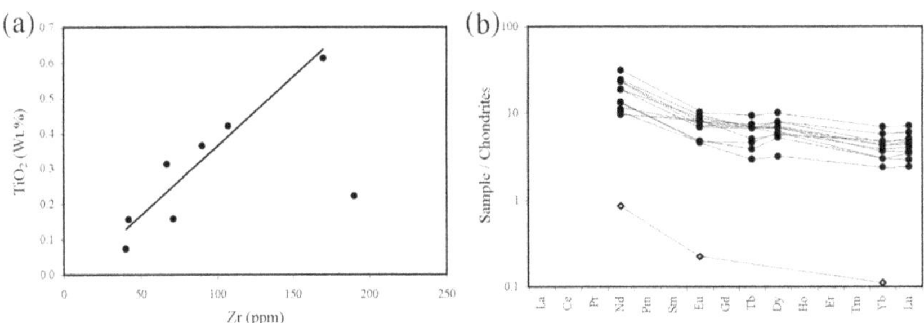

Figure 4.2:
(a) Covariation of Zr with TiO_2 in the analysed beach sands. The trendline (r = 0.948) is calculated after removal of sand IT85 from the dataset. (b) Chondrite-normalised rare earth element patterns of the sands, glasses and shell analysed. (d) Comparison of Sr concentrations as obtained via ICP–OES and INAA. Black dots are the sand samples and the open diamond the shell material. Reprinted from Archaeometry, Vol 56, Brems, D., Degryse, P., Trace element analysis in provenancing Roman glass-making, 116-136, Copyright (2014), with permission from John Wiley and Sons.

Concentrations of Rb range from 23.6 to 38.9 ppm in the analysed sands. A correlation between K_2O and Rb (r = 0.847) indicates that Rb is present in K-feldspar. Cs values range from 0.34 to 4.26 ppm, with sands IT34 and SP20 having the highest values. There is no clear correlation between Cs and the other alkali elements. Ba concentrations in the sands vary between 95 and 280 ppm. Ba is often said to be related to alkali feldspar. However, the data show no correlation between Ba and K_2O (r = 0.182) or between Ba and Rb (r = -0.077). Although not very strong, correlations are present between Ba and Sr (r = 0.816) and Ba and Ca (r = 0.642). This suggests that Ba is not exclusively related to feldspars and that it can also be derived from the carbonate fraction of the beach sands. However, the shell material analysed in this study only contains 15 ppm Ba. A connection to barite could not be investigated since concentrations of S are not available.

Ga could only be detected in one sample, SP22, where it reaches 13 ppm. Detection limits for this element were, however, generally of the same order of magnitude. Concentrations of Th range from 1.55 to 5.45 ppm in the sands. Th is correlated with Zr (r = 0.839), Hf (r = 0.838) and the (heavy) rare earth elements (r = 0.726 – 0.739), indicating its presence in zircon (Wang et al., 2011; Nardi et al., 2012).

Rare earth element concentrations are chondrite-normalised using the values of Sun and McDonough (1989) and the REE patterns are shown in Fig. 4.2b. Only Nd, Eu, Tb, Dy, Yb and Lu could be analysed. Ce was below the detection limit in all samples. REE patterns of the sands are relatively flat, with a slight enrichment in light rare earth elements (LREE). Only sand sample FR17 shows a small positive Eu anomaly, consistent with its relatively high feldspar content (Brems et al., 2012b, c). For the other samples analysed no significant Eu anomalies were found. REE concentrations in the shell fragments are more than one order of magnitude lower than in the sands. The chondrite-normalised REE pattern of the shell material is similar to those of the sands, although possibly somewhat more enriched in LREE (Fig. 4.2b).

4.2.3   Lime-related elements

The concentration of Sr in the sands analysed varies between 32 and 315 ppm with a good correlation between Sr and Ca (r = 0.892). The use of Sr isotope ratios and Sr concentrations for provenancing Roman natron glass, and especially the source of lime used, is extensively discussed in Chapter 3 and in Brems et al. (2013a).

Shell fragments have relatively low concentrations of most of the trace elements analysed. Only Sr is strongly concentrated in shell (1550 ppm). Cl and Br are present in concentrations similar to those found in the sands. The other elements analysed occur in concentrations at least one order of magnitude smaller than in the sand samples. Results for the REE concentrations are comparable to those reported for clam shells from the North Sea by Wedepohl et al. (2011b).

4.2.4   (De)colourant-related elements

Mn and Sb are often present in elevated concentrations in Roman natron glass (Sayre and Smith, 1961; Sayre, 1963; Henderson, 1985; Jackson, 2005; Freestone, 2008; Silvestri, 2008; Foster and Jackson, 2009). These elements were deliberately added to the batch to make the glass colourless by oxidising iron, or to combine with other elements to create a variety of colours. The MnO concentration in the analysed sands ranges from 0.01 to 0.11%. These values can be seen as the background level for Mn, which can be attributed to the sand raw material (see also Brems et al., 2012b, c). Concentrations of MnO higher than 0.1% in natron glass are influenced by deliberate addition or by recycling. As an impurity in the Mn decolourising agent, extra Ba is often introduced to the glass resulting in a strong positive correlation between the two elements (Brill, 1988; Jackson, 2005; Silvestri, 2008). The data presented in this study shows no significant correlation between MnO and Ba (r = 0.285). Sb contents of more than 37,500 ppm (4.99% Sb_2O_5) have been reported for natron glass (Arletti et al., 2006). Most of the sands analysed in this study contain low concentrations of Sb, i.e., below 1.4 ppm. Only sand FR16 has higher Sb levels of 19.2 ppm. Zn concentrations of most sands analysed vary between 5.3 and 84 ppm. Sand sample FR16, however, contains 272 ppm of Zn. The very high Zn and Sb content in sand FR16 can be attributed to the former exploitation of a Pb-Zn-Ag deposit in Les Bormettes (Féraud, 1983; Artignan and Nauchbaur, 2007). Exploitation of this small scale ore deposit already commenced during the Gallo-Roman period and ended in 1908 (Artignan and Nauchbaur, 2007). Erosion and redeposition of material from local tailings results in relatively high concentrations of metals in the local beach sand. Next to Zn and Sb, soils in the area locally contain elevated concentrations of Pb, Cd, Cu and Hg (Artignan and Nauchbaur, 2007).

Of all sands analysed, only SP22 contains measurable amounts of Ni, i.e., 21 ppm. Measured concentrations of Co in the sands vary between 1.1 and 4.5 ppm. Cu, Se and Hg concentrations were below the INAA detection limit for all samples analysed. Ag is below the detection limit for all samples except for sand IT85 which contains 0.80 ppm Ag. In is only above the detection limit in sand sample SP20, where it reaches 0.03 ppm.

4.3   Trace elements as a provenance indicator for the silica source?

Most of the suitable Roman-type sand raw materials from the western Mediterranean can be distinguished from those of the east based on their Nd isotopic composition. Sand raw materials and raw natron glass from Egypt and Syro-Palestine have relatively homogeneous ε_{Nd} values between -6 and -4 (Degryse and Schneider, 2008; Freestone et al., nd; this study). Western Mediterranean beach sands mostly have lower ε_{Nd} values, i.e., usually lower than -8. However, in Chapter 3, we have seen that Roman-type sand raw materials from two locations in the southeast of Italy also have relatively high ε_{Nd} values, which may coincide with those thought to be typical for raw glass produced in the eastern Mediterranean. Differences in trace element patterns may help to resolve this problem.

To evaluate the use of trace elements as a provenance indicator for Roman natron glass, we must compare the trace elemental signature of the sands analysed in this study to those of the known glass groups and raw materials from the eastern Mediterranean. For easier comparison, trace element data are often normalised to a common reference. In glass studies, this is usually the average continental earth's crust (Freestone et al., 2002a; Wedepohl et al., 2011a, b). In this study, we use the average continental crust values of Wedepohl (1995).

Selected trace element data for raw natron glass and glass vessels from the Byzantine-Islamic primary workshops at Bet Eli'ezer and Apollonia on the coast of Israel are given by Freestone et al. (2000). Their compositional profiles (Fig. 4.3a) are relatively uniform, indicating the geochemical homogeneity of the sand sources along the Levantine coast. Average trace element patterns for Belus River sands are indeed very similar in shape (Fig. 4.3b; Brill, 1988, 1999; Brems and Degryse, 2014). Glass from a workshop at Tel el-Ashmunein, Egypt, shows a very different trace element distribution (Fig. 4.3c). They are readily distinguished from the Levantine glasses by their lower Sr and Ba, and higher Zr concentrations (Freestone et al., 2000). The different origins of these glass groups was confirmed by Sr isotopic analysis, which suggested the use of shell-bearing beach sands for the production of the Levantine glasses, and inland sand and limestone for the Egyptian glass (Freestone et al., 2003). Trace elemental compositions similar to those of the Levantine raw glass, were also found in glasses from Cyprus and Anglo-Saxon England (Freestone et al., 2002a, 2008b). Other glasses, for example from Carthage, have both high Sr and Zr, indicating that their primary origin lies elsewhere (Freestone et al., 2000).

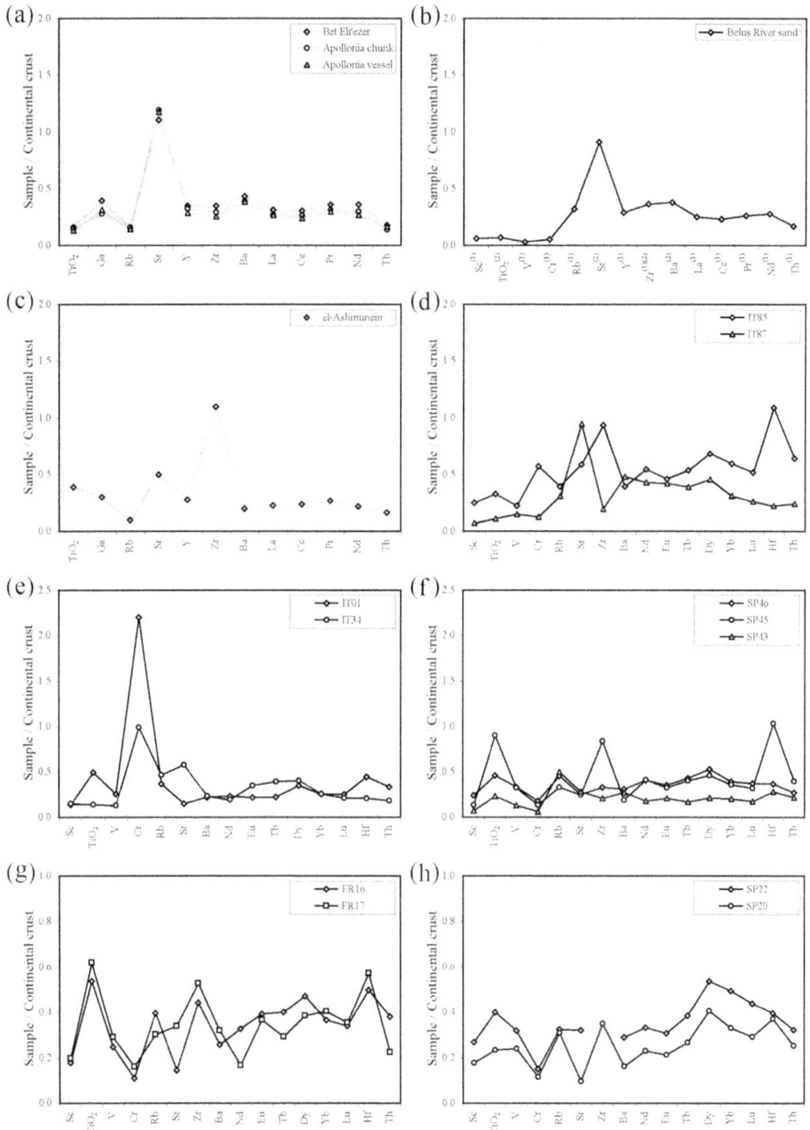

Figure 4.3:
Trace element concentrations normalised to the mean abundances in the earth's continental crust (Wedepohl, 1995). Black lines are sand samples, grey lines represent glass samples. (a) Levantine raw glass chunks from Bet Eli'ezer and Apollonia, and glass vessels from Apollonia (Freestone et al., 2000); (b) Belus River sand ([1] Brems and Degryse, 2014; [2] Brill, 1988, 1999); (c) Egyptian glass from Tell el-Ashmunein (Freestone et al., 2000); (d) Sand samples IT85 and IT87; (e) Sand samples IT01 and IT34; (f) Sand samples SP46, SP45 and SP43; (g) Sand samples FR16 and FR17; (h) Sand samples SP20 and SP22. Reprinted from Archaeometry, Vol 56, Brems, D., Degryse, P., Trace element analysis in provenancing Roman glass-making, 116-136, Copyright (2014), with permission from John Wiley and Sons.

Sand raw material IT87 has an ε_{Nd} value of -4.17 (Chapter 3; Brems et al., 2013b). This is on the higher end of the ε_{Nd} values of -6 to -4 characteristic of raw natron glass from Israel (Degryse and Schneider, 2008; Freestone et al., nd). However, ε_{Nd} values as high as -1 have been reported for sands 400 m south of the mouth of the Belus River (Degryse and Schneider, 2008) and Nile River sediments (Goldstein et al., 1984; Weldeab et al., 2002; Scrivner et al., 2004). Therefore, it is possible that such a suitable sand source with high ε_{Nd} values also exists along the Syro-Palestinian coast. Sand raw material IT85 has an ε_{Nd} value of -6.11 (Brems et al., 2013b). Trace element signatures of these Italian sands are shown in Fig. 4.3d. The two trace element patterns are markedly different. Sand IT87 generally has the lower concentration of trace elements. Only Sr and Ba are relatively elevated. Concentrations of TiO_2, Cr and Zr are very low. Sand sample IT85 generally contains higher concentrations of heavy minerals and the associated trace elements TiO_2, Cr, Zr, Hf and REE. When comparing Fig. 4.3a, b and d, the trace element pattern of sand IT87 appears to be very similar to those of Syro-Palestinian raw glass and Belus River sands. However, IT87 is even more depleted in Zr. Another important difference lies with the characteristically low Al_2O_3 content of sand IT87, which resulted in Al_2O_3 contents lower than 1.5% in the glass (see Brems et al., 2012b, c). This is significantly lower than the Al_2O_3 concentrations of 2.2 – 3.2% in Belus sand (Brill, 1988, 1999) and 2.5 – 4.0% generally found in the Syro-Palestinian glass made from that sand (Freestone et al., 2000). Sand raw material IT85 can be easily distinguished from Belus River sand by its higher TiO_2, Cr and Zr contents, and lower Sr. Sand IT85 has $^{87}Sr/^{86}Sr$ isotope ratios higher than the modern seawater value, while Egyptian glass has relatively low Sr isotopic signatures (Freestone et al., 2003; Brems et al., 2013a; Fig. 4.4).

Other Roman-type sand raw materials along the coasts of the western Mediterranean Sea all have significantly lower ε_{Nd} values (Brems et al., 2013b). These sand sources can be further separated based on their ε_{Nd} values and trace element signatures (Fig. 4.4). Italian sands IT01 and IT34 have similar Nd isotopic signatures as suitable (low lime) sand raw materials from the southwest of Spain (SP46, SP45 and SP43). ε_{Nd} values of these sands all lie between -9.40 and -7.99 (Brems et al., 2013b). IT01 and IT34 can be distinguished from each other by their different TiO_2 and Cr contents (Fig. 4.3e). Sand IT01 is also much lower in Sr, a result of the very low calcium carbonate content of this sand. After the addition of extra shell fragments as a source of lime, glasses produced from these two sands would have similar Sr concentrations (Brems et al., 2013a). Trace element signatures of the Spanish sands vary according to their heavy mineral contents (Fig. 4.3f). Sample SP43 is the most pure quartz sand and has very low concentrations of for example TiO_2, Zr, Hf and REE. Sands SP46 and especially

SP45 contain more heavy minerals and have relatively elevated TiO_2, Zr, Hf and REE. The most distinguishing feature between the Italian and Spanish sands is the Cr content. The Italian sand samples contain on average 12 times more Cr than the Spanish sands.

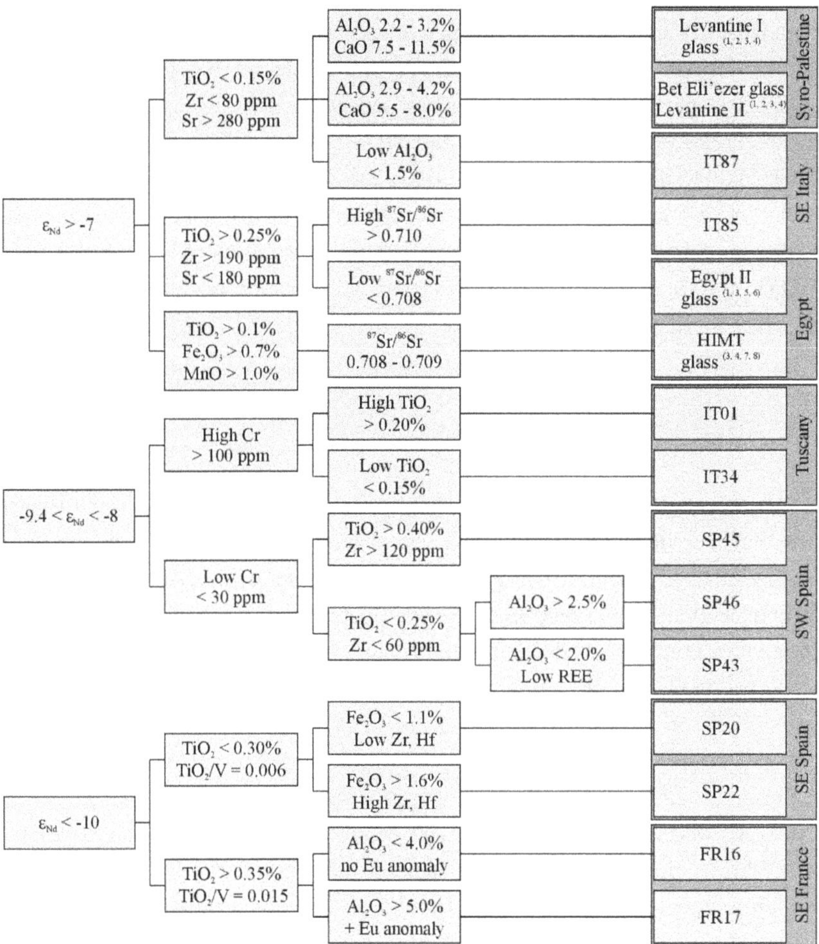

Figure 4.4:
Flowchart for determining the provenance of natron glass. [1] Freestone et al., 2000; [2] Freestone, 2006; [3] Degryse and Schneider, 2008; [4] Freestone et al., nd; [5] Gratuze and Barrandon, 1990; [6] Freestone et al., 2003; [7] Freestone et al., 2005; [8] Foster and Jackson, 2009. Reprinted from Archaeometry, Vol 56, Brems, D., Degryse, P., Trace element analysis in provenancing Roman glass-making, 116-136, Copyright (2014), with permission from John Wiley and Sons.

Roman-type glassmaking sand raw materials from southeastern Spain (SP22 and SP20) and southeastern France (FR16 and FR17) all have ε_{Nd} values lower than -10 (Brems et al., 2013b). Trace element signatures of the French sands are shown in Fig. 4.3g. Sand sample FR17 has slightly higher concentrations of most trace elements. Rb and Th, however, are higher in FR16 and also the REE pattern is somewhat different with an enrichment of Eu in FR17. Sand samples SP22 and SP20 have trace element patterns with very similar shapes (Fig. 4.3h). The pattern of SP22 is generally shifted to higher concentrations. This sand also contains relatively elevated Fe_2O_3 and Al_2O_3 levels. SP20 shows two pronounced peaks for Zr and Hf, indicating a relative enrichment of the heavy mineral zircon. These peaks are absent in the SP22 trace element pattern. The two sand samples from the southeast of France contain on average higher concentrations of TiO_2 and also the TiO_2/V ratio is different with 0.014 – 0.016 for French sands and 0.006 – 0.009 for Spanish sands.

4.4   Conclusions

In this chapter, we have evaluated the use of trace elements as a provenance indicator for Roman natron glass. Sand raw materials from the western Mediterranean, suitable for Roman natron glass production, were analysed for their trace elemental compositions and compared to raw materials and raw glass from known primary production centres in the east.

It was shown that the combined use of Nd isotopic signatures, major elements (particularly Al_2O_3) and trace element patterns makes it possible to distinguish between the different possible sources of suitable sand raw materials in the regions under investigation. Trace elements that proved to be the most diagnostic are Ti, Cr, Sr, Zr and Ba. Apart from Ba (and possibly Sr), which is often associated with Mn decolourants, these elements are seldom influenced by the addition of colouring agents or recycling, and should provide direct information about the nature of the silica source used. However, since data for possible sand sources from areas such as North Africa, Greece, Turkey and Cyprus are limited or not yet available, the existence of competing Roman glass producers with overlapping elemental and isotopic characteristics in these areas can neither be excluded nor confirmed.

Slightly elevated concentrations in glass of trace elements commonly associated with colouring agents, such as Mn, Co, Ni, Cu, Zn, Sb and Pb, are often interpreted as the result of recycling of glass cullet. The analysis of these elements in Roman-type glassmaking sands provides a good idea of the background levels that can be attributed to impurities in the source of silica. The current dataset suggests that for the two most commonly used decolourisers, MnO and Sb, these background levels are 0.1% and 30 ppm respectively. The presence of higher amounts of these elements in Roman glass would indicate their deliberate or accidental (due to recycling of cullet) addition. Additional detailed analysis of a wider range of trace elements in suitable sand raw materials can only provide further insights into the influence of the different raw materials on the composition of ancient natron glass. Also, trace elemental data of materials typically used as colourants in antiquity is essential to evaluate their contribution to the final trace elemental composition of the glass. Especially their influence on the concentration of elements commonly attributed to the sand source should be investigated to make sure that these elements are indeed only derived from the sand raw material and are therefore potentially useful as provenance indicators. To our knowledge, this kind of data is not available at the moment.

Chapter 5

The Sources of Natron

V. Devulder, P. Degryse

Ancient glass consists of three main components: a silica source as a glass network former, a stabilizer in the form of lime, and a flux. This flux is needed to lower the melting temperature of the silica (Freestone, 2006). Plant ashes or mineral natron were added as such a melting aid. Glasses made with plant ash can be distinguished from natron glass based on chemical analysis. Low magnesia/potash (< 1.5 wt%) glass is made with natron, while high magnesia/potash glass is made with plant ashes as a fluxing agent (Sayre and Smith, 1961). Natron glass is widespread in the Mediterranean area between the second half of the first millennium BC and the ninth century AD. Before and after, glass made with plant ash is common.

In this chapter, the focus of research lies on the provenance of the natron used as a flux in Greco-Roman glass making. Natron, also referred to as *natrun* (Shortland, 2004), is actually a mixture of different minerals formed in the evaporation of alkaline lakes (Fig.5.1). These deposits contain minerals such as natron ($Na_2CO_3.10H_2O$), trona ($Na_2CO_3.NaHCO_3.2H_2O$), burkeite ($Na_6CO_3.2SO_4$) and halite (NaCl), amongst others, in varying proportions. The origin of the natron (whereby we here use the term to indicate a mixture of minerals) used for Greco-Roman glass production is a matter of debate. Several possible sources of natron have been suggested by several ancient authors.

Strabo mentions in his Geography *"Above Momemphis are two nitre-beds, which contain very large quantities of nitre"* (Strabo *Geography* 17.1.23; Jones, 1932: 73). This might refer to the Wadi Natrun in Egypt, then consisting of two lakes, but it has also been suggested that this passage could refer to the Wadi Natrun as one deposit, and al-Barnuj as another (Lucas and Harris, 1962). There is no evidence, however, that the ancient Egyptians knew al-Barnuj, and whether there was natron harvested there (Currie, 2008).

Pliny writes in his Natural History *"The soda-beds of Egypt used to be confined to the regions around Naucratis and Memphis, the beds around Memphis being inferior. For the soda becomes stone-like in heaps there, and many of the soda piles there are for the same reason quite rocky"* (Pliny *NH* 31.46; Jones, 1963:

447). According to Lucas and Harris (1962), al Barnuj is near Naucratis and consequently, the Wadi Natrun is then near Memphis (although it is physically not that near).

Pliny also writes in his Natural History "*At Clitae in Macedonia is found in abundance the best, called soda of Chalestra, white and pure, very like salt. There is an alkaline lake there with a little spring of fresh water rising up in the centre. Soda forms in it about the rising of the Dog-star for nine days, ceases for nine days, comes to the top again and then ceases*" (Pliny *NH* 31.46; Jones 1963: 445). This is interpreted by Ignatiadou et al. (2005) and Dotsika et al. (2009) as being Lake Pikrolimni in Greece. However, this location is not mentioned in Pliny's chapter on glass making (book XXXVI, chapter 66), where he discusses the various kinds of glass and the modes of making it.

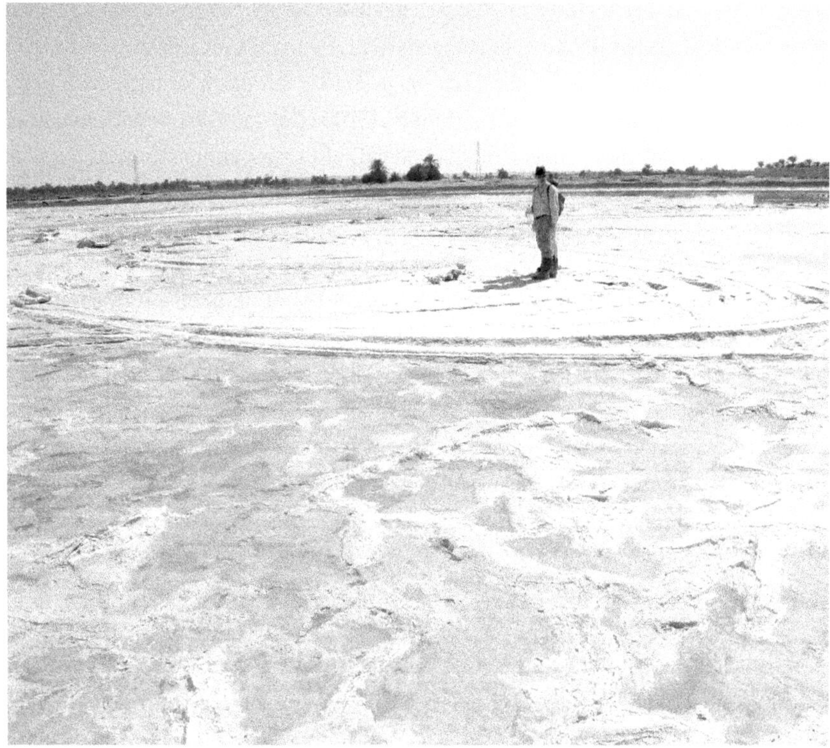

Figure 5.1
Photograph of present day deposits of burkeite and halite at Wadi el Natrun (picture taken September 2006). Picture reprinted from Degryse, P., Scott, R.B, Brems, D., The archaeometry of ancient glassmaking: reconstructing ancient technology and the trade of raw materials, *Perspective* (2014-2), INHA.

Despite geochemical analysis, until now, no scientific evidence existed for the use of either one or multiple sources of natron in ancient glass making. Moreover, finding natron *in situ* nowadays has proven difficult, as the lakes mentioned here precipitate mainly halite and sulphate minerals, rather than forming carbonates. Moreover, most of northern Africa, the most suitable environment for saline lakes, remains unexplored. One such previously unstudied deposit of natron in Fazzan (Libya) was kindly brought to our attention by David Mattingly (University of Leicester), while another evaporitic deposit near Sebket al Melah (Tunisia) was sampled by Dennis Braekmans (Leiden University). These evaporites will be discussed here as well.

5.1   Provenancing natron

To provenance the flux used in glass making, B (boron) is a promising element. It enters ancient glass mainly through the flux, and the B isotopic composition of minerals can vary due to natural isotope fractionation. As a consequence, different geological sources of sodium carbonate can have different B isotopic signatures (it can span a range of -20 to +40‰ $\delta^{11}B$) and the B isotope ratio could be used to differentiate multiple sources of natron flux. A method for the isolation and measurement of B in glass and its raw materials was developed (Devulder et al. 2013a, b, 2014) and used within the ARCHGLASS project.

A set of 50 natron glasses consisting of primary glasses and secondary/recycled glasses were analysed for their B content and isotopic composition (Table 5.2). The primary glasses date between the 6th and 8th century AD and originate from Israel (the glass 'factories' of Bet-Eli'ezer and Apollonia). The secondary and recycled glasses date between the 4th century BC and the 9th century AD and originate from Tel el-Ashmunein (Egypt), Kelemantia (Slovakia), Grandcourt Farm (UK), Oudenburg (Belgium) and Sagalassos (Turkey). The colours of the glass samples are diverse, ranging from natural green-aqua to colourless. The B concentration in the secondary and recycled glass shows B concentrations between 100 and 200 µg/g, whereas the primary glasses show lower values, ranging from 30 to 50 µg/g. The $\delta^{11}B$ values of the secondary and recycled glasses are homogeneous and range between +27 and +33‰ while the primary glasses show slightly lower $\delta^{11}B$ values.

Natron sources from Egypt (Wadi Natrun and al-Barnuj), Libya (Fezzan), Tunisia (Sebket al Melah) and Greece (Lake Pikrolimni) were also investigated for their B signature (Table 5.1). All these deposits were sampled during the 21st century AD and rarely contain natron or trona as the major mineral phase. Most of them contain halite (NaCl) and burkeite ($Na_6CO_3.2SO_4$). For an optimal study

Sample	Country	Location	$\delta^{11}B$ (‰)	[B] ($\mu g \cdot g^{-1}$)	XRD
WN 3	Egypt, Wadi Natrun	Lake Hamra, southwestern side	28,6	6,2	**Halite**
WN 11	Egypt, Wadi Natrun	Lake Zug, southwestern side	29,4	46,6	**Halite**
WN 12	Egypt, Wadi Natrun	Lake Zug	29,0	59,3	**Halite**, [trona, burkeite]
WN 15	Egypt, Wadi Natrun	Lake Beida, northeastern side, edge	24,3	7,0	**Halite**
WN20	Egypt, Wadi Natrun	Lake Beida, northeastern side	34,3	62,6	**Halite**, [burkeite]
WN 23	Egypt, Wadi Natrun	Lake Fazda, western side	30,9	20,3	**Halite**, burkeite
WN 28	Egypt, Wadi Natrun	Lake Fazda	30,5	25,5	**Halite, burkeite**, [trona, gypsum]
WN30A	Egypt, Wadi Natrun	Lake Fazda, western side, 200m from edge	31	21,7	**Trona**, [burkeite, halite]
WN 30B	Egypt, Wadi Natrun	Lake Fazda, western side, 200m from edge	30,9	36,9	**Burkeite, halite**, [trona]
WN33	Egypt, Wadi Natrun	Lake Hamra, southwestern side	31,2	179,1	**Burkeite, halite**
WN 36b	Egypt, Wadi Natrun	Lake Zug, southwestern side	27,6	9,6	**Burkeite, halite**, [trona]
L1	Libya	Fezzan	27,1	179,4	**Halite, trona & quartz**, [burkeite, aphthitalite]
L2	Libya	Fezzan	28,1	45,2	**Trona, burkeite**, thermonatrite, nahcolite, quartz
L3	Libya	Fezzan	19,4	41,0	**Trona, halite, quartz**, burkeite, aphthitalite
L4	Libya	Fezzan	26,4	373,4	**Trona, thermonatrite, halite, quartz**, [burkeite & aphthitalite]
T3	Tunisia	Sebket al Melah	32,2	4,7	**Halite**
T4	Tunisia	Sebket al Melah	31,5	9,5	**Halite**
T5	Tunisia	Sebket al Melah	31,9	7,8	**Halite**
Pikro	Greece	Lake Pikrolimni	10,6	38,3	**Halite**, [burkeite]
Pikro2	Greece	Lake Pikrolimni	10,4	182,3	**Halite**, [thenardite]
al-Barnuj	Egypt	al-Barnuj	28,5	11,6	**Halite, trona**, [burkeite, thenardite]

*Bold = major, normal = minor, [brackets]= trace

Sample	Country	Location	$\delta^{11}B$ (‰)	[B] ($\mu g \cdot g^{-1}$)	εNd
SP20	Spain	El Rubial	-4,0	15,0	-11,76
SP22	Spain	Las Marinas de Vera (Carrucha)	-2,2	25,8	-11,33
SP43	Spain	Sanlucar de Barrameda	-3,8	11,0	-9,20
SP45	Spain	Mazagon	-10,3	63,6	-8,68
SP46	Spain	Isla Canela	-9,8	41,1	-7,99
FR16	France	Les Bormettes, La Londe-les-Maures	-10,5	23,4	-12,40
FR17	France	Cavalaire-Sur-Mer	-11,4	47,4	-10,14
IT85	Italy	Metaponto Lido	-6,2	11,8	-6,11
IT34	Italy	Torre del Sale, Piombino	-5,0	14,7	-9,40
IT87	Italy	Masseria Maime	-3,3	8,5	-4,17
EG6	Egypt	Makadi Bay	+10,7	5,1	nd
ALEX	Egypt	Alexandria	+22,5	10,6	bdl
Li09.01	Libya	Great Sand Sea	-7,6	7,8	nd
Li09.02	Libya	Erg Murzuq	-5,1	3,6	nd
GR1	Greece	Acharavi, Korfu	-1,9	15,9	-6,51
TUR1	Turkey	Gümbet	-3,4	21,5	-7,40
TNIO	Tunisia	Tunis	-6,7	15,1	-11,00

Table 5.1
Mineralogy and B isotopic composition of present day natron and sand deposits (salt samples from Wadi el Natrun sampled and described by Shortland, 2004; Nd isotopic data of the sand deposits analysed by Brems, 2012)

of glass making in ancient times, samples from ancient context with natron/trona as main minerals should be analysed. The B concentration values of the salts analysed range from 5 to 400 µg/g. Their $\delta^{11}B$ values show a wide range. The samples from Egypt range between +25 and +35‰ with an average of +29‰. The al Barnuj sample is indistinguishable from the Wadi Natrun samples. The samples from Fezzan mostly show $\delta^{11}B$ values of around +25‰, very similar to the Egyptian sources. Also the Sebket al Melah samples range are very similar to the Wadi Natrun material, averaging +32‰ $\delta^{11}B$. The Pikrolimni samples have a much lower $\delta^{11}B$ of +10 ‰. No correlation was found between the $\delta^{11}B$ value and the mineral phases present.

Sample	Colour	Date	Place found	$\delta^{11}B$ (‰)	[B] (ug.g^{-1})	εNd
SA-2007-VL-358	Colourless	Second half 2nd century	Sagalassos	30,1	195,3	-6,3
SA-2007-VL-1161	Yellow	1st -7th century AD	Sagalassos	32,6	253,1	
SA-2007-VL-744	Colourless	1st -7th century AD	Sagalassos	29,7	139,1	
SA-2007-VL-678	Colourless	1st -7th century AD	Sagalassos	30,2	161,4	
SA-2007-VL-219	Colourless	1st half 4th century AD	Sagalassos	28,7	162,1	
SA-2007-VL-135	Colourless	2nd half 4th century, 1st half 5th century AD	Sagalassos	28,7	149,9	
SA-2007-VL-539	Yellow/green	1st -7th century AD	Sagalassos	29,4	237,9	
SA-2007-VL-115	Colourless	1st -7th century AD	Sagalassos	29,7	148	
SA-2007-VL-88	Colourless	1st -7th century AD	Sagalassos	28,4	160,3	
SA-2007-VL-211	Colourless	1st -7th century AD	Sagalassos	27,3	151,2	
SA-2007-VL-304	Pale yellow	1st -7th century AD	Sagalassos	27,1	158,4	
SA-2006-VL-7	Colourless	End 5th century AD/ 6th century AD	Sagalassos	28,8	210,1	-5,2
SA-2007-VL-26	Dark blue	1st -7th century AD	Sagalassos	30,1	162,9	
SA-2005-VL-31	Brown	1st -7th century AD	Sagalassos	31	170,6	
SA-2007-VL-1167	Pale green	1st -7th century AD	Sagalassos	27,1	176,9	
SA-2007-VL-1085	Colourless	1st -7th century AD	Sagalassos	27,7	166,5	
SA-2007-VL-1095	Yellow	1st -7th century AD	Sagalassos	32,2	290,6	
8933	Pale blue/ green	4th – 5th century AD	Oudenburg	28,7	159,6	
8926 C	Pale blue/ green	3rd century AD	Oudenburg	29,7	136,3	
71310	Pale blue	3rd century AD	Oudenburg	28,2	154,1	-5,6
23993	Pale blue/ green	4th -5th century AD	Oudenburg	28,7	159,6	-6,1
2960	Pale blue	4th -5th century AD	Oudenburg	27,4	193,7	-6,5
SA-2006-VL-05	Yellow/green	1st -7th century AD	Sagalassos	27,4	264,7	
SA-2007-VL-136	HIMT	1st -7th century AD	Sagalassos	26,3	192,9	
SA-2007-VL-137	HIMT	1st -7th century AD	Sagalassos	27,7	198,6	
SA-2007-VL-138	HIMT	1st -7th century AD	Sagalassos	26,9	187,6	
SA-2007-VL-205	HIMT	1st -7th century AD	Sagalassos	29	218,8	
KEL2	Pale blue	between 175 and 179 AD	Kelemantia	-4,4	31,1	-7,3
KEL3	Pale yellow	between 175 and 179 AD	Kelemantia	29	108,3	-6,1
KEL4	Colourless	between 175 and 179 AD	Kelemantia	29,9	152,7	-9,0
GF1	Dark yellow	4th- 1st century BC	Grandcourt Farm	31,6	227,7	-4,0
GF4	Colourless	4th- 1st century BC	Grandcourt Farm	34	101,3	-4,0
GF5	Dark blue	4th- 1st century BC	Grandcourt Farm	30,5	109,4	-12,0
BE36	Blue-green	6th-8th century AD	Bet Ell'ezer	25,1	25,3	-4,7

Sample	Colour	Date	Place found	δ^{11}B (‰)	[B] (ug.g^{-1})	εNd
BE37	Blue-green	6th-8th century AD	Bet Eli'ezer	27,2	31,3	-4,8
BE38	Blue-green	6th-8th century AD	Bet Eli'ezer	28,0	41,4	-4,8
BE39B	Green	6th-8th century AD	Bet Eli'ezer	26,8	34,5	-5,0
BE42	Yellow	6th-8th century AD	Bet Eli'ezer	24,6	31,0	-5,1
BE44	Yellow	6th-8th century AD	Bet Eli'ezer	28,6	35,7	-4,7
AP4	Olive	6th-7th century AD	Apollonia	26,2	39,4	-4,6
AP5	Aqua	6th-7th century AD	Apollonia	29,9	48,2	-4,5
AP6	Aqua	6th-7th century AD	Apollonia	27,0	28,3	-4,4
AP8	Aqua	6th-7th century AD	Apollonia	22,7	35,1	-4,1
AP9	Aqua	6th-7th century AD	Apollonia	23,8	36,6	-4,3
AP13	Olive	6th-7th century AD	Apollonia	25,7	38,9	-4,4
AP14	Olive	6th-7th century AD	Apollonia	25,8	44,3	-4,3
AP15	Olive	6th-7th century AD	Apollonia	25,1	35,9	-4,6
TA1	Pale blue	8th-9th century AD	Tel el-Ashmunein	24,9	77,8	-7,0
TA2	Pale green	8th-9th century AD	Tel el-Ashmunein	25,2	75,1	-6,5
TA3	Pale blue	8th-9th century AD	Tel el-Ashmunein	24,7	87,5	-6,9
TA4	Pale blue	8th-9th century AD	Tel el-Ashmunein	20,4	60,5	-6,1

Table 5.2
B isotopic composition of ancient natron glass (samples previously described by Lauwers (2008) for Sagalassos, Ganio et al. (2012b, c) for Oudenburg, Freestone et al. (2003) for Tel el-Ashmunein, Freestone et al. (2000) for Bet Eli'ezer, Tal et al. (2004) for Apollonia, Degryse et al. (2009b) for Kelemantia).

Sands from around the Mediterranean area (Spain, France, Italy, Greece, Tunesia, Egypt, Turkey and Libya) were likewise analysed, as possible contributors to the B balance of an ancient glass. The sands investigated have an average B concentration of 18 µg/g, ranging from 3 to 60 µg/g, indicating the silica source is not insignificant in contributing to the B balance in an ancient natron glass. No correlation was found between the geographical origin and the B content of the sand. All sands show heavily negative δ^{11}B values from -10 to 0‰ (except for 2 very lime rich sands from Egypt with a δ^{11}B of +10 and +20‰). Consequently, through the contribution of the silica source, the δ^{11}B value of an ancient glass will always be lower than that of the flux used.

5.2   The origin of ancient natron

Most North African salts analyzed (Wadi Natrun, al-Barnuj, Fezzan and Sebket al Melah) show high $\delta^{11}B$ values similar to Greco-Roman glasses. The north African sources can moreover not be distinguished from each other by B isotopic analysis, which is not surprising in the light of their very similar precipitation environment and geological background. Some differences have been noted, however, in the trace element composition of the salts, especially in their potassium contents (ranging from a few ppm up to several percent in the Fezzan natron; Fabri and Degryse, 2013). Conversely, the Lake Pikrolimni material shows a much lower $\delta^{11}B$ value (+10‰), not in accordance with the ancient glasses in this study. Under the assumption that the $\delta^{11}B$ value of the mineral raw materials did not change over time, and that the $\delta^{11}B$ of ancient glass is in between that of the flux and the silica source, in the current state of research lake Pikrolimni can be excluded as a possible natron source in Roman glass making.

No significant difference in $\delta^{11}B$ is observed between the glasses originating from different regions or different time periods. This might suggest that, through time and space, one source of flux was used, or multiple sources with a similar $\delta^{11}B$. The contribution of B from the sand to the total amount of B in the glass complicates the picture for provenance determination of the flux. Not only the B concentrations and $\delta^{11}B$ of flux and sand will affect the $\delta^{11}B$ value of the glass, also the ratio flux/sand plays an important role. The lower $\delta^{11}B$ of the primary glasses might suggest a different source of natron, however such difference may also be explained by the smaller influence of the $\delta^{11}B$ of the flux on the final composition. The primary glass analysed here dates to a later period, when less natron was used in glass making. This means less B entered the glass from the flux source, and the influence of the sand source is relatively greater, making the glass depleted in $\delta^{11}B$.

All these variables make it difficult to assess whether one or multiple sources of flux were used. It is clear however, that the $\delta^{11}B$ value of Lake Pikrolimni is inconsistent with its use as a flux source in antiquity. It is also clear that all ancient natron glass analysed here, is very homogenous in B isotopic composition. The fact that all natron sources from north Africa analysed so far are very similar (to identical) in isotopic composition, and consistent with ancient natron glass, makes placing the source of all flux for natron glass making in this wider area very tempting. The recent discovery of natron deposits in Fezzan, shows that there may be many such sources yet undiscovered in this part of the Roman world. The high potassium content of this particular deposit, could also be a promising feature to distinguish possible sources of natron in glass. A large range in the K_2O content of

Roman natron glass throughout the empire has been noted before, and remained unexplained until now, but could effectively be due to the use of flux sources rich in this element. The occurrence and mining of natron throughout north Africa would certainly make the supply of this raw material across the Mediterranean to primary factories in Italy of the western provinces much easier, possible even with overland transport.

Since the B isotopic analysis of Roman glass and flux sources has just started, many aspects still need to be addressed. The analysis of ancient natron would be of great interest, especially to investigate whether present-day natron is characterized by a $\delta^{11}B$ value similar to its ancient equivalent. However, with the results obtained so far, the Wadi Natrun (and by extension North Africa) as sources of natron for Roman glass making is still very probable.

Chapter 6

Primary glass factories around the Mediterranean

P. Degryse, M. Ganio, S. Boyen, A. Blomme, B. Scott,
D. Brems, M. Carremans, J. Honings, T. Fenn, F. Cattin

Glass objects and chunk glass were sampled from consumer and secondary production sites as well as shipwrecks and primary factories. In total almost 350 samples of glass from 30 archaeological excavations were analysed geochemically and compared to the signature of possible sand raw materials and known primary glass factories. The chronology of the samples was determined by stratigraphical association. The samples come from securely dated contexts between the 6th century BC (possibly going into the 8th century BC) and the 8th century AD (possibly going into the 9th century AD) and are all natron glass. Plant ash glasses (with a MgO/K_2O content >1.5 wt%) do occur even within this timeframe, but were not taken into account in this discussion. Rosenow and Rehren (2014) have suggested an Egyptian origin for the particular type of dark green unguentaria made in plant ash glass that were also present in our sample set.

The aim of the comparison was to establish a large scale diachronic image of the primary origin of natron glass in the Hellenistic-Roman to early Byzantine/ Islamic world. Only material for which major and trace element contents and Sr-Nd isotope ratios are known, are included in the discussion presented in this chapter. Analytical data can be found in full in digital form on http://ees.kuleuven. be/geology/archaeometry/index.html. As indicated before, the reconstruction of the primary origin of the glass presented here refers to broad geological/ geographical regions where glass was (likely) made, rather than to individual sites. Our reference data on north Africa remain limited in the light of the political situation there during the course of the ARCHGLASS project, prohibiting field work.

6.1   The corpus of glass

Glass was sampled from sites across the extents of the Mediterranean world and the Roman Empire.

The far east side of this territory is accounted for by glass from archaeological sites at Petra, Tannur, Barsinia in Jordan, Jerusalem and Shechem in Israel, Tel el Ashmunien in Egypt, Sagalassos in Turkey and Pieria in Greece. Also the primary glass from Bet Eli'ezer and Apollonia comes from this area. The northern part of the territory is represented by glass from Tienen and Oudenburg in Belgium, Gonio and Pichvnari in Georgia, Bocholtz and Maastricht in the Netherlands, Grandcourt Farm in the UK and Kelemantia in Slovakia. In the western part of the area studied, glass was sampled from Barcino in Spain and sites in France, including the Embiez shipwreck. The central part of our study area comprises Italy, with glass sampled at Augusta Praetori, Herculaneum, Pompeii and Potentia, including also the Iulia Felix shipwreck. Glass from northern Africa was limited to material from Carthage in Tunisia.

Some samples were individual samples with little excavation context, such as material from Iron age France (kindly provided by B. Gratuze) or from the Harvard Semitic Museum and Harvard Art Museums (kindly provided by J. Greene, A. Aja and S. Ebbinghaus). However, most samples have an extensive excavation context, and previously published data.

6.1.1   The East

The Pieria samples are from ancient Pydna (present day Makrygialos) and its environs. The city itself flourished during the 5th to 4th century BC, and in the Roman period. Extensive cemeteries have been excavated, mainly to the north of Makrygialos, and to the south in Kitros Alykes and Kitros Louloudia. Some further samples were found at ancient Methone, a city built in the 8th century BC and destroyed in the 4th century BC. Excavated parts of the town include a 12 m deep cellar belonging to the 8th to 7th century BC, the Agora dated to the 6th to 4th century BC and a contemprorary Acropolis.

The glass from Sagalassos (SW Turkey) represents the common colour varieties of window and free-blown vessel glass and originates from Roman to Byzantine excavation contexts (Poblome, 1999; Degeest, 2000). Glass from three distinct periods was sampled: imperial (1st to 3rd century AD), late Roman (4th to first half 5th century AD) and early Byzantine (second half 5th to 7th century AD) (Degryse et al., 2009b). Most samples are from the third period, reflecting the majority of the deposits hitherto excavated at Sagalassos (Degryse et al., 2005). Both colourless and naturally coloured glass was sampled.

The site of Barsinia is situated in north-west Jordan, 15 km west of Irbid. This site was occupied continuously from the Hellenistic until the Abbasid Period (332 BC to 950 AD). It contains domestic architectural remains next to several tombs and a number of ceramic kilns. Glass samples Ba1 and Ba10 were found in section D1, sample Ba3 in section B9. All these sections of the site are dated to late Roman times (3rd to 4th century AD; Savage and Keller, 2007; Savage et al., 2008), while sample Ba8a was dated to the 5th to 6th century AD.

From the ancient city of Petra, located in south-west Jordan, colourless glass fragments dated to the 3rd to 5th century AD were analysed. Petra played an important role in Arabian trade during Nabataean, Roman and Byzantine times. Several excavation campaigns conducted over several years yielded a multitude of glass vessels, objects and window glass which have been studied both typologically and chemically (Marii, 2001; O'Hea, 2001; Schibille et al., 2008). The majority of the glass samples analysed come from the 'Great Temple' in the city centre, and were retrieved from the temple's large forecourt, the Lower Temenos' which, like the temple, was abandoned some time from the late 5th to mid 6th century AD (Joukowsky, 2007). This glass assemblage is dated to the 4th to early 5th centuries AD (Schibille et al., 2012).

The ruins of the Nabataean pilgrimage sanctuary of Khirbet et-Tannur are located on a hilltop, 7 km north of the ancient village of Khirbet edh-Dharih, the third caravan stop north of the Nabataean metropolis Petra on the traditional north-south route, the Kings Highway. The main phase of the surviving temple complex of Tannur, which was constructed in the first half of the 2nd century AD, received further additions and repairs in the 3rd century AD. The pottery and lamps show that religious activities, with burnt offerings and dining, continued there through the 3rd and into the 4th century AD, with the earthquake of 363 AD providing a *terminus ante quem* for the colourless glass samples studied here (Schibille et al. 2012).

The samples from Bet Eli'ezer (near Hadera) and Apollonia (near Arsuf) are naturally aqua to green coloured primary glasses, made in tank furnaces in Israel (Freestone et al., 2000; Tal et al., 2004). Tel el-Ashmunein is a secondary workshop in Egypt where the raw primary glass was shaped and coloured (Freestone et al., 2003). The samples from Bet Eli'ezer are dated between the 6th and the 8th century AD, those from Apollonia between the 6th and the 7th century AD and the Tel el-Ashmunein samples are dated between 8th and the 9th century AD.

6.1.2   North Africa

Excavations in the "Precinct of Tanit", the so-called "Tophet" of Carthage, Tunisia, by the American Schools of Oriental Research Punic Project between 1976 and 1979 under the direction of Professor L.E.Stager yielded cremated remains of human infants, and in some cases young animals, buried in cinerary urns (Stager, 1980, 1982; Stager and Wolff, 1984; Stager et al., forthcoming). The excavated materials and records are currently housed at the Semitic Museum, Harvard University, and contained a large number of glass beads, dating to the 8th to 4th century BC, including rare monochrome blue glass beads, small numbers of blue and white eye beads, some red beads with white and yellow spots and large numbers of black and brownish beads (Eremin et al., 2011).

6.1.3   The North

The Pichvnari necropolis on the Black Sea coast of Georgia lies in an area known in the late 1st millennium as 'Colchis'. The burials of the necropolis date to the late 5th century BC and frequently contain grave goods, including the core-formed strong coloured vessel glass analysed here (Shortland and Schroeder, 2009).

The Grandcourt Farm samples are Iron Age glass eye beads dated to the 2nd to 1st century BC. They were excavated at Grandcourt Quarry in East Winch, Norfolk by Archaeological Project Services for SIBELCO UK (Malone, 2010). They were dated by association and radiocarbon dating of the layers they were found in.

The Roman town of Tienen, founded in the 1st century AD, is situated in Belgium, close to the ancient road from Cologne to Boulogne, as part of the *civitas Tungrorum* (Cosyns, 2001-2002; Cosyns and Martens, 2002-2003). Archaeological excavations proved the production of several goods such as pottery, iron, and bronze as well as (secondary) glass vessels. The samples discussed are all aqua vessel glass fragments, associated with a 2nd century AD secondary glass furnace.

Oudenburg was a Roman castellum situated on the Belgian coast with direct access to the North Sea and a connection to the Roman road network. The castellum started out in the 1st century AD as a small trade settlement with a small port but expanded in the 2nd century AD. In the 3rd century AD the civil settlement was abandoned and the focus became strictly military (Vanhoutte, 2008; Vanhoutte et al., 2009). The glass samples from Oudenburg, all naturally coloured, were collected during excavations of the south-western corner of the castellum. Two samples are dated to the 3rd century AD, others belong to the 4th to 5th century AD.

The Roman auxiliary fort of Iža (Kelemantia) in Slovakia is situated about 4 km east of the confluence of the rivers Waag and Danube. The remains of the earth-and-timber fortification all belong to one single construction phase dating between 175 and 179 AD (Degryse et al., 2009b). The wooden construction was laid to waste by German attackers or was dismantled, abandoned and set on fire by the Roman forces themselves when they left in 179 AD. Under Emperor Commodus' rule, a stone *castellum* was built on exactly the same location, occupied until the end of the reign of Valentinianus I in 375 AD, when barbarians invaded the frontier zone. Samples KEL1, KEL2, KEL3 and KEL5 belong to excavation layers of the earth-and-timber camp, dated to 175-179 AD. Sample KEL4 comes from excavation layers in the *castellum* and was dated to the 3rd century AD (Degryse and Schneider, 2008; Degryse et al., 2009b).

The glass from Bocholtz (the Netherlands) originates from an underground burial chamber in a known Roman graveyard (de Groot, 2006). The chamber was dated to the last quarter of the 2nd to the first quarter of the 3rd century AD. Glass grave goods were identified and sampled for analysis. Sample BO 106 is free-blown colourless plate with a greenish tinge, Isings type 42b (Isings, 1957), dated to the 2nd century AD. Sample BO 109 is a free-blown colourless cylindrical bottle, Isings type 51b (Isings, 1957), dated to the late 2nd to early 3rd century AD. Sample BO 123 is a colourless cast or slumped small bowl.

Gonio, ancient Asparos, is a fortified city located in south-western Georgia, 15 km south of Batumi. The fortress was occupied from the 1st to 4th century AD, strategically placed upon the crossroads of the roads to the east (Turkey) and to the south (Armenia) (Kakhidze, 2008). All samples are naturally coloured glass.

Maastricht (the Netherlands) sample Ma3a was retrieved from a grave in the Scharnweg in 1986. Besides pottery, several glass objects such as beakers and bowls were recovered. Typochronologically all the material can be dated to the first half of the 3rd century AD (Panhuysen and Dijkman, 1987, p.212 and afb.11). Sample M5a was excavated in 1983 under the Hotel Derlon, in layers assigned to the second quarter of the 5th century AD (Dijkman, 1993, Fig.9-C1 en D8). Both samples are blue (aqua) coloured.

6.1.4   The West

Samples from the Embiez shipwreck are colourless raw glass chunks and colourless window panes and cups. The Embiez, wrecked off Embiez island, in southern France, is dated to the end of the 2nd to the beginning of the 3rd century AD (Foy and Fontaine, 2007). The ship was about 12 m long, and is at the moment the only Roman ship known totally dedicated to the trade of glass. The majority of the cargo was composed of raw glass, and is estimated to be c. 18 tons. 163 kg of this has been recovered, sub-divided into 65 blocks of different size. In addition, the cargo was composed of about 1800 cup fragments, and window glass panels.

Sant Boi de Llobregat is a site situated in the province of Barcelona. Architectural remains were found which indicate continuous occupation from the 6th century BC until the 12th century AD. It is located 17 km from the sea on the bank of the river Llobregat, close to the important port of Tarraco. Romans occupied the city from the 1st to 5th century AD, and recently there have been some indications near this site for the secondary production of Roman natron glass. Both naturally coloured and colourless Roman glass samples, dated 1st to 5th century AD, were analyzed.

6.1.5   The central Mediterranean

The famous eruption of mount Vesuvius on August 23rd 79 AD was responsible for the destruction of the ancient cities of Pompeii and Herculaneum, covering both towns in volcanic ash, pumice and lava (Horne, 1895). Pompeii as a town was an important passage way for goods that arrived by sea and had to be sent towards Rome or Southern Italy along the nearby Appian Way. Its nice location and climate, and its numerous sources of recreation, soon attracted the wealthier Romans.

The Pompeii samples were provided by the Museo Archeologico Nazionale di Napoli. Unfortunately the precise proveniences of these glass fragments, all naturally colourless or intentionally decoloured, are not known since they were stored in the museum deposits and precise data about the archaeological context were not available. Moreover, the fragmentary state of the samples did not allow for typological characterization. The Herculaneum samples were provided by the Getty Conservation Institute, through Prof. Giacomo Chiari, thanks to a collaboration with the Herculaneum Conservation Project and Dr. Domenico Camardo. These glass fragments consist of five colourless, naturally or intentionally decoloured, three yellow, two emerald green, three blue and one green-blue. These samples were excavated in the *Insula Orientalis II* sewer (Camardo, 2007). As with the Pompeii samples, the fragmentary state of the Herculaneum samples did not allow for typological characterization.

Augusta Praetoria (modern Aosta, north-west Italy) was founded in 25 BC following the Roman conquest of a territory previously inhabited by the indigenous Salassi. The town commanded the access to the passes allowing communication between the south and the north of the Alps. It was situated on the main route of the Gallic Consular Road. It connected the eastern roman towns of Italy (Aquileia and Adria) with the western Italian towns of Augusta Praetoria and Augusta Taurinorum, and the northern provinces on the other side of the Alps. The glass fragments analysed are dated to between the 1st and the 4th century AD. Both colourless and naturally coloured vessel glass fragments occur.

Samples from the Iulia Felix wreck represent colourless glass, typologically identified as bottles, cups and plates. The Iulia Felix was found off the coast of Grado, in the north-eastern part of Italy, and has been dated to the first half of the 3rd century AD (Toniolo, 2003; Silvestri, 2008; Silvestri et al., 2008). It was a small cargo ship, approximately 15 to 18 m long, of the so-called corbita type. The cargo was mainly composed of amphorae of various types, together with more than 11000 fragments of glass, totaling 140 kg. The original shapes of the different glass artefacts can be deduced from a typological study of the fragments. They all represent examples of accidentally broken glass, most probably collected for recycling (Toniolo, 2003).

Potentia lies south of the modern town of Porto Recanati (MC), about 100 m inland from the Adriatic coast on the ridge of a beach near the ancient mouth of the river Potenza. Livius, in the *Ab Urbe Condita* books, reported the founding of the coastal colony of Potentia in 184 BC. Official sources for the later history of the town are, however, minimal. Epigraphic evidence testifies to the flourishing development of the town from the Augustan Age onward into the late 2nd century AD. The latest finds belong to the 7th century AD but the exact character of the occupation in the town at this period is unclear. Overall, the position and layout of the town reflects that of typical maritime colonies along the Tyrrhenian coast, built as bridgeheads for land and sea routes and tied to the roads Via Flaminia and Via Salaria (Vermeulen and Verhoeven, 2006; Vermeulen et al., 2006). The 50 glass samples analysed here were excavated from different contexts during archaeological campaigns carried out by Gent University under the supervision of Prof. Frank Vermeulen in the period 2000-2010. Both colourless and naturally coloured glass from different periods is represented.

Previously, analytical data (elemental and isotopic) of the Embiez and Iulia Felix glass have been published in Ganio et al. (2012a, b, c) and Ganio (2013). Data on Augusta Praetoria, Potentia, Herculaneum and Pompeij were presented in Ganio (2013). Analytical data on Petra, Barsinia, Gonio, Augusta Praetoria, Sant Boi de Llobregat, Oudenburg and Tienen were published by Ganio et al. (2012a, b). Data on Sagalassos, Maastricht, Bocholtz and Kelemantia were published in Degryse and Schneider (2008) and Degryse et al. (2005, 2009a, b). Analytical data on the samples from Petra and Tannur were published by Schibille et al. (2012). Major element and Sr isotopic data and on primary glasses from Bet Eli'ezer, Apollonia and on Tel el-Ashmunein have been published by Freestone et al. (2000), Tal et al. (2004) and Freestone et al. (2003). Major element data on the Carthage samples were presented by Eremin et al. (2011), major and trace element data on the Pichvnari glasses by Shortland and Schroeder (2009).

6.2   Provenance indicators

From the study of sand deposits around the Mediterranean, and from major and trace element and Sr-Nd isotope ratio analysis of ancient primary glass, the most important chemical provenance indicators of a primary glass origin can be derived (see chapters 3 and 4). In terms of major to trace element composition, the Al_2O_3, TiO_2 and Zr contents of ancient glass are the most relevant indicators for a differing primary origin of the silica raw material used. In terms of isotopic analysis, the Nd isotopic composition of ancient glass is able to distinguish between different silica sources used. Combined, these chemical indicators have the most potential to source primary glass making in the Greco-Roman world to different geological-geographical areas around the Mediterranean basin. No primary factories outside the eastern Mediterranean, however, are known from archaeological excavations.

Only a few regions and geological situations so far have been recognized as having suitable silica sands for making natron glass with a Greco-Roman composition. Sands from Israel and Egypt are known as an eastern Mediterranean source of glass. Sands from Spain, France and Tuscany in Italy have been described as possible western Mediterranean silica sources for ancient glass making. Sands from Apulia and Basilicata in the heel of Italy can be described as the possible central Mediterranean or Italian sources of silica sand for glass making. Only a limited study of sands from North Africa was possible, in view of the geopolitical situation there are the time of this work. However, suitable sands seem to be present in Libya and Tunisia next to Syro-Palestine and Egypt.

When silica sands from the eastern Mediterranean are used for glass making, the ε_{Nd} signature of the resulting glass is less negative than ε_{Nd} -7. E.g. beach sands from near the mouth of the river Belus show high Nd isotopic signatures between -1 and -4.8 ε_{Nd}. Syro-Palestinian glass from primary factories such as those in Bet Eli'ezer or Apollonia, usually termed Levantine I or II glass, has an Al_2O_3 content higher than 2 wt%, a TiO_2 content lower than 0.1 wt% and a Zr content lower than 80 ppm. Its Sr-Nd isotopic signature is between $^{87}Sr/^{86}Sr$ 0.7088 to 0.7092 and ε_{Nd} -4.1 to -6.0. Egyptian glasses, termed Egypt I and II, likewise have an Al_2O_3 content higher than 2 wt%, a TiO_2 content higher than 0.25 wt% and a Zr content higher than 190 ppm. Their Sr isotopic signature is usually lower than $^{87}Sr/^{86}Sr$ 0.7086. Their ε_{Nd} signatures are unknown. HIMT glass, presumed to have been made in Egypt (Nenna, 2014), has an Al_2O_3 content higher than 2 wt%, a TiO_2 content higher than 0.1 wt% and an Fe_2O_3 and MnO content respectively higher than 0.7 and 1.0 wt%. Its Sr isotopic signature is usually between $^{87}Sr/^{86}Sr$ 0.7075 and 0.7090, its Nd isotopic signature is between -4 and -6 ε_{Nd}.

When silica sources from the western Mediterranean would be used for glass making, the ε_{Nd} signature of the resulting glass is expected to be more negative than ε_{Nd} -7. Sands from Tuscany and southwest Spain would give ancient glass an intermediate ε_{Nd} signature between -8 and -10. Sands from Tuscany can be distinguished by their high Cr content (> 100 ppm), while the Spanish sands have much lower Cr content (< 30 ppm). Sands from France and the southeast of Spain would give ancient glass a signature of ε_{Nd} lower than -10. The French sands can be distinguished from the southeastern Spanish sands by their higher TiO_2 content of higher than 0.35 wt%.

When silica sources from Apulia and Basilicata in the heel of Italy (also: the central Mediterranean) would be used for glass making, the ε_{Nd} signature of the resulting glass will be less negative than ε_{Nd} -6, while its Al_2O_3 content would be lower than 2 wt% and its TiO_2 and Zr content would be respectively lower than 0.1 wt% and lower than 80 ppm.

For North African glass, admittedly our database is still very limited. Most sands from the Tunisian to Western Egyptian dessert approach the ε_{Nd} signature of inland Saharan sand, with an Nd isotopic signature between -9 and -14 ε_{Nd} for sands sampled from the Great Sand Sea in Egypt and Libya.

One special sand that needs to be discussed is a single sample from Alexandria (sample EG08DB05 in Appendix A). It is very high in lime content with 30wt% CaO, and has an Al_2O_3 content of 0.3 wt%, a TiO_2 content of 0.03 wt% and a Zr content of 20 ppm, with an ε_{Nd} -3.99. Though the sand is unsuitable to make a Roman compositional type glass, it is interesting to observe the occurrence of this type of pure sand with low minor and trace element contents in this area. It could as such not be distinguished from sands from Apulia and Basilicata, using the here defined provenance indicators. It is clear more work is needed to construct an extensive database of Syro-Palestinian to north African sands to establish a firm background of the isotopic signatures of suitable glass making sands in the eastern Mediterranean.

It has to be stressed that none of these indicators are insensitive to glass mixing, and recycling of glass. When primary glass sources are mixed, be it as primary glass or in recycling cullet, all these chemical indicators will show an intermediate content in all parameters in the resulting glass. Therefore, in the analysis of the dataset presented here, extreme values for all these indicators were looked for. When a possible sand raw material for glass making has the lowest or highest content in one of the chemical indicators used here, finding such a low/ high content in the glass dataset is extremely relevant. It is the perfect indicator for the fact that this source material has been used in making primary glass, and that its chemical signature has been preserved in natron glass after several centuries, despite recycling and mixing.

It is assumed that recycling of glass or the use of cullet will affect the trace element content of a glass batch and can as such be detected. Usually, an elevated Pb content is regarded as indicative of glass recycling. High lead glasses (such as opaque glass, strong coloured glass or mirror glass) may end up accidentally in a glass batch, and will affect the Pb content of the resulting glass in a very significant way, raising levels of Pb in the glass to over 100 ppm. Most sand sources analysed here have a low Pb content (< 100 ppm), supporting the theory that a Pb content > 100 ppm in Roman glass is due to either colouring/opacifying or recycling.

6.3   Provenance of Greco-Roman natron glass

Looking at the analytical dataset (Fig. 6.1), 16 samples or 5% of the glass assemblage shows an $\varepsilon_{Nd} < -7$, indicative of a primary origin in the western Mediterranean. These glasses are all naturally coloured or colourless, and come from 11 sites all over the sampling area. They are dated to the 4th century BC and between the 1st century AD and the first half of the 5th century AD. This is only a limited and chronologically well defined part of the dataset. Some of these glasses show an $^{87}Sr/^{86}Sr > 0.7092$.

Most of the glass in our dataset, 210 samples or 64% of the glass assemblage, shows an $\varepsilon_{Nd} > -7$ with an Al_2O_3 content of higher than 2%, indicative of a primary origin in the eastern Mediterranean. This glass is naturally coloured as well as colourless or strong coloured, is found on all sites studied here and hence occurs over the entire time period studied. About a third of these samples have a Pb content of higher than 100 ppm, and appear to be recycled glass. The very early strong coloured glass from Carthage, Pieria, Pichvnaro and Shechem shows Pb contents of higher than 1000 ppm, likely related to the colourants and opacifiers used for this glass.

However, very significantly, 72 samples or 22% of the glass assemblage, show an $\varepsilon_{Nd} > -7$ with an Al_2O_3 content lower than 2%, indicative of a primary origin not in the known Syro-Palestinian or Egyptian factories but possibly in the heel of Italy, or an unknown (African) source. This group is found on many sites all over the sampling area, but most samples come from Augusta Praetoria and from the Embiez and Iulia Felix shipwrecks, next to Carthage. Almost all of the Italian glass is colourless (mostly Sb decoloured), can be dated between the 1st and the 4th century AD (Fig. 6.2.), and has a low Pb content, showing no indication of recycling or mixing. An exceptional group of Sb decoloured glass has an extremely high Pb content of several thousand ppm, likely related to the antimony mineral used in its manufacture. A group of strong coloured (mainly black) glass with less negative ε_{Nd} and low Al_2O_3 content originated from Carthage and can be dated to the 4th to 5th century BC. This group again has very high Pb contents of over 1000 ppm, likely related to the colourant. Rosenow and Rehren (2014) used the predominance of 1st to 2nd century Sb decoloured glass at ancient Bubastis (northern Egypt) to suggest a closeby primary origin of this glass in Egypt, though no sands matching such Roman glass composition have been identified in this study.

25 samples or 7% of the glass assemblage shows an $-7 < \varepsilon_{Nd} < -6$, not immediately indicative of a specific primary origin, but an intermediate value between all sand sources studied. 15 of these samples have a Pb content of over

100 ppm, which could be indicative of their recycled nature. The samples are naturally coloured as well as colourless or strong coloured.

Less than 2% of our dataset is glass with an origin likely in Egypt. This is not surprising, as HIMT glass, easily recognizable by its colour, has been avoided in this study (only sampled in Sagalassos), and as Egypt I and II glass has rarely been found outside of Egypt itself. Remarkably, one colourless sample from Petra does show the typical composition of Egypt I glass.

Most of the glass (90%) in the dataset shows an $^{87}Sr/^{86}Sr$ between 0.7087 and 0.7091 with several hundreds of ppm Sr in the glass, indicative of shell as a lime source in the glass. This justifies the sampling of beach sand for this study.

Almost a quarter (23%) of the total glass dataset are samples with a lead content between 100 and 1000 ppm Pb. While this shows that much of the dataset will have been recycled, this figure is still rather limited. This is promising for the possibility of looking at this dataset in terms of primary provenance of silica raw materials and glass origins. Around 10% of the total dataset are colourless or strong coloured samples with a Pb content higher than 1000 ppm. While an association of the element Pb with opacifiers and strong colourants or decolourants is not surprising, it is remarkable that a group of colourless glass with such high Pb contents exists. This is particularly interesting in the light of mixed glass batches and Pb as an indicator for recycling, as remelting such glass in a low-Pb glass batch, would increase the total Pb content of the final recycled material significantly, without affecting its colour or properties.

In terms of glass with a non-eastern Mediterranean origin, the largest spread in glass composition and hence the location of primary glass factories is clearly situated in the early Roman period, between the 1st and 4th century AD (Fig. 6.1). The large spread in the isotopic composition of this Roman glass, suggests primary production of glass all over the Mediterranean, but is also likely smeared out between the extreme values of the dataset because of intense recycling of glass with different primary origins and thus different signatures. The earliest natron glass with a likely non-eastern Mediterranean origin can be found in Carthage and is dated to the 4th century BC.

Hypothetically, if three primary production centres from the 1st to 4th century AD were each using their own local sand raw material, a Syro-Palestinian source would have a ε_{Nd} of -6, an Italian factory would have an ε_{Nd} of -4 and a western Mediterranean source would have an ε_{Nd} of -8 or -9. If the glass from these three factories was not mixed or recycled together with the glass from other sources, three distinct glass groups would appear, each one showing its own Nd isotopic signature. However, as recycling was a common process (clear from our dataset with a quarter of the glass having Pb contents higher than 100 ppm), the isotopic

	BC 600-550	BC 550-500	BC 500-450	BC 450-400	BC 400-350	BC 350-300	BC 300-250	BC 250-200	BC 200-150	BC 150-100	BC 100-50	BC 50-0	AD 0-50
-3,5													
-3,75													
-4								1					5
-4,25								2					17
-4,5			1	1	5	5	1	1	1	1	1	1	14
-4,75	3	3	3	3	6	6	2	2	1	1			12
-5	1	1	6	6	5	5					1	1	17
-5,25	1	1	6	6	2	2					1	1	23
-5,5													11
-5,75	1	1	1	1	1	1							2
-6													6
-6,25	2	2	2	2									1
-6,5													1
-6,75													4
-7					1	1							1
-7,25													
-7,5													
-7,75					1	1							
-8													
-8,25													1
-8,5													
-8,75													
-9													
-9,25													
-9,5													
-9,75													
-10													
-10,25													
-10,5													
-10,75													
-11													
-11,25													
-11,5													
-11,75													
-12													
-12,25													
-12,5											1	1	1
-12,75													
-13													
	600-550	550-500	500-450	450-400	400-350	350-300	300-250	250-200	200-150	150-100	100-50	50-0	0-50
	BC	BC	BC	BC	BC	BC	BC	BC	BC	BC	BC	BC	AD

AD	AD	AD	AD	AD	AD	AD	AD	AD	AD	AD	AD	AD	AD	AD
50-100	100-150	150-200	200-250	250-300	300-350	350-400	400-450	450-500	500-550	550-600	600-650	650-700	700-750	750-800
5	3	4	4	2			1	1	1	1				
17	16	15	14	14					1	1	1	1		
14	10	12	24	13	12	11	4	3	11	11	11	11		
12	9	12	20	10	7	7			6	6	6	6	3	3
17	11	10	19	13	10	9	5	4	9	9	9	9	5	5
23	13	18	25	14	13	13	4	3	3	3	3	3	1	1
11	3	9	13	2	3	3	3	1						
2	2	3	7	2	3	3	3	4	3	3	3	3		
6	4	5	5	4	3	2								
1		2			1	1	1	1					1	1
1	1		1										1	1
4		2	1		1	1	1	1					1	1
1	1	1	1										1	1
	1	2											1	1
	1	1												
1	1	2	1											
			2	2	1	1	2							
	1	1												
	1	1												
1														
					1	1	1							
50-100	100-150	150-200	200-250	250-300	300-350	350-400	400-450	450-500	500-550	550-600	600-650	650-700	700-750	750-800
AD	AD	AD	AD	AD	AD	AD	AD	AD	AD	AD	AD	AD	AD	AD

Figure 6.1:
Representation of the occurrence of ε_{Nd} signatures (grouped per quarter unit) through time (grouped per 50 years) for the full dataset of natron glass analysed. Sample chronology was determined by stratigraphical association. Samples are counted in full for each time block of their assumed date. Colours represent the total count for each time interval: green is 1 to 3, yellow 4 to 5, orange 6 to 9 and red higher than 9.

	BC 600-550	BC 550-500	BC 500-450	BC 450-400	BC 400-350	BC 350-300	BC 300-250	BC 250-200	BC 200-150	BC 150-100	BC 100-50	BC 50-0	AD 0-50
-3,5													
-3,75													1
-4													
-4,25					1	1							2
-4,5					3	3							4
-4,75			1	1	3	3							8
-5					1	1					1	1	9
-5,25													5
-5,5													2
-5,75													3
-6													2
-6,25													1
-6,5													3
-6,75													
-7													
-7,25													
-7,5													
-7,75													
-8													
-8,25													
-8,5													
-8,75													
-9													
-9,25													
-9,5													
-9,75													
-10													
-10,25													
-10,5													
-10,75													
-11													
-11,25													
-11,5													
-11,75													
-12													
-12,25													
-12,5													
-12,75													
-13													
	600-550	550-500	500-450	450-400	400-350	350-300	300-250	250-200	200-150	150-100	100-50	50-0	0-50
	BC	BC	BC	BC	BC	BC	BC	BC	BC	BC	BC	BC	AD

AD 50-100	AD 100-150	AD 150-200	AD 200-250	AD 250-300	AD 300-350	AD 350-400	AD 400-450	AD 450-500	AD 500-550	AD 550-600	AD 600-650	AD 650-700	AD 700-750	AD 750-800
	1	1	1	1	1	1								
1	1	2	2	1	1	1								
	2	2	2	2	2	2								
2	5	7	16	8	7	6	2	1	1	1	1	1		
4	4	9	9	6	3	3								
8	6	6	9	9	5	4	1							
9	6	12	18	9	7	7	2	1						
5	1	7	9	1	4	4	3	1						
2	2	3	6	1	3	3	3	1						
3	2	2	2	1	1									
2	1	2	1	1										
1	1		1											
3		2	1											
								1	1					1
		1	1											1
														1
			2	2	1	1	1	2						
					1	1	1							
AD 50-100	AD 100-150	AD 150-200	AD 200-250	AD 250-300	AD 300-350	AD 350-400	AD 400-450	AD 450-500	AD 500-550	AD 550-600	AD 600-650	AD 650-700	AD 700-750	AD 750-800

Figure 6.2:
Representation of the occurrence of ε_{Nd} signatures (grouped per quarter unit) through time (grouped per 50 years) for the colourless glass analysed. Sample chronology was determined by stratigraphical association. Samples are counted in full for each time block of their assumed date. Colours represent the total count for each time interval: green is 1 to 2, yellow 3 to 4, orange 5 to 7 and red higher than 7.

signatures of these factories would be smeared out between the highest and lowest values with progressive recycling. In this case, the intermediate signature of the Syro-Palestian workshop would be obscured. When the Italian and western Mediterranean workshops stop making glass in the 4th to 5th century AD, due to any possible reason, their signatures would not disappear abruptly but would die out slowly as recycling continues but only Syro-Palestinian primary glass remains to be made, resulting in a gradual narrowing of the isotopic range back to the Syro-Palestinian signature. The length of the period in which the signature of the former glass producers remains visible in the glass record is difficult to estimate and would depend on the intensity of recycling, the ratio between recycling of old glass and new raw production, the concentration of Nd in the different glass types, etc.

It is clear that the most ubiquitous signature in this dataset, for the entire time span studied from maximally the 8th century BC to the 9th century AD, is that of an eastern Mediterranean sand source and points toward continuous natron glass making in Syro-Palestine and/or Egypt for nearly 1500 years, and recycling of this glass throughout that time span.

Chapter 7

Conclusions

P. Degryse

The main goals of the ARCHGLASS project were to develop archaeometrical techniques to reconstruct ancient economies, and to apply these methods to gain insights into the trade and processing of mineral raw materials used for glass making. The project further developed Sr-Nd isotopic techniques and newly invented B isotopic techniques as a means to characterize glass and this book presents a database of mineralogical and chemical compositions of possible raw materials (sand and flux) for primary glass making. From these data, the primary provenance of ancient natron glass, with focus on the Hellenistic-Roman world, can be derived.

In particular, the occurrence of primary production centres of raw glass outside those known from archaeological excavation in Syro-Palestine and Egypt were investigated. The western Mediterranean area including the regions of Italy, Gaul, Spain (described by ancient authors as primary glass producers) and north-Africa were investigated. The political situation in Northern Africa and the Middle East during the course of the project prohibited much fieldwork there. This makes the study of North Africa as a supplier of natron glass more difficult. Moreover, a database of analyses of possible raw materials and glass is never complete. The approach to the geological materials analysed here is based on regional survey, taking samples from different geological hinterlands with a wide spacing, but not going into very detailed sampling grids. The reconstruction of the primary origin of the glass presented here therefore refers to broad geological/geographical regions where glass was (likely) made, rather than to individual sites. Our reference data on north Africa remain limited.

It is clear from our survey that suitable glass making sands are relatively rare. Six limited areas outside the eastern Mediterranean could be defined where suitable sand raw materials would have been available to the Roman glassmaker:

(1) Beach sand between the mouth of the River Basento and the River Bradano (Basilicata Region, SE Italy) and (2) in the area southeast of Brindisi, on the northeastern side of the Salentina peninsula (Puglia Region, SE Italy), would

produce glass with properties close to Roman glass, but with a Al_2O_3 concentration generally lower than that of typical natron glass.

(3) Sands from the western part of the Follonica Gulf, between Piombino and Follonica (Tuscany Region), are also suitable for glass production. These sands are only suitable after the addition of extra calcium carbonate.

(4) Beach sands near the mouth of the Rio Guadiana (Huelva Province, SW Spain), (5) near the town of Águilas in the Murcia Region (SE Spain), and (6) from the Bay of Hyère (Département du Var, Provence, SE France) are all very rich in quartz and contain only very small amounts of shell or limestone fragments. After the addition of extra $CaCO_3$, these sands would produce glass with a typical Roman composition.

These results of course do not prove that there was a Roman primary glass production industry in the western Mediterranean, but demonstrate that, if so, the suggested regions are the most likely suppliers of silica raw materials in the western Mediterranean.

Sr and Nd isotopic signatures were assessed as a technique to provenance natron glass. Binary mixing equations showed that the $^{87}Sr/^{86}Sr$ isotope ratio of glass is not always indicative of the main source of lime. Sands from the west contain important amounts of radiogenic Sr, resulting in a shift in $^{87}Sr/^{86}Sr$ ratios to higher values in the glass. Since the Sr isotopic signatures of Nile-dominated sediments in the eastern Mediterranean lie around or slightly below the modern-day seawater signature, the induced shift in the Sr isotopic signature of the glass there is less explicit. For glasses produced in the eastern Mediterranean, the Sr isotopic signature is indeed a good provenance indicator for the source of lime. In the western part of the Mediterranean, this is only true for glasses with low Al_2O_3 concentrations and an Al_2O_3/CaO ratio lower than 0.25. The Sr content of the glass appears to be a better and more robust indication of the lime source.

Nd in Roman natron glass originates from the non-quartz (heavy) mineral fraction of the sand raw material and its isotopic composition is an indication for the source of the silica. Suitable glassmaking sands from Spain, France and the western part of Italy all have relatively low values in Nd isotopic signature between -12.0 and -7.0 ε_{Nd} and glass produced from them could be readily distinguished from glass from the known primary production sites in Egypt and Syro-Palestine. Two good sand sources in the Basilicata and Apulia Regions (SE Italy), however, have ε_{Nd} values higher than -6 ε_{Nd}, which may coincide with those of glass with an eastern Mediterranean origin. These eastern Mediterranean beach sands were sampled by R.H. Brill in the sixties (Brill 1999) and analysed by Degryse and Schneider (2008) and Brems et al. (2012a, b) in the course of this project. They show high Nd isotopic signatures between -1 and -4.8 ε_{Nd}. Primary glass from the factories at Bet Eli'ezer and Apollonia, using Israeli coastal sand (Brill, 1999) and active between the 6th and 8th century AD, produce raw glass with an isotopic signature between -4.1 en -5.1 ε_{Nd}. Freestone et al. (nd) earlier reported values for Bet Eli'ezer, Apollonia and HIMT primary glass between -5.0 and -6.0 ε_{Nd}.

A Nd isotopic signature of ε_{Nd} lower than -7.0 seems a justified cut off for the primary origin of glass not lying in the eastern Mediterranean. Moreover, the combined use of Nd isotopic signatures, major elements (particularly Al_2O_3) and trace element patterns makes it possible to distinguish between the different possible sources of suitable sand raw materials in all the regions under investigation. Glass with ε_{Nd} higher than -6.0 and with elevated Al_2O_3 and/or Zr/TiO_2 has a primary origin in the eastern Mediterranean, while glass with ε_{Nd} higher than -6.0 and Al_2O_3 lower than 2wt% seems to originate either from the Italian peninsula, or a yet unknown source in northern Africa (where a low Al_2O_3, Zr and Ti_2O sand has been identified near Alexandria). In effect, as data for possible sand sources from North Africa and some islands of the Mediterranean are still limited, the existence of competing Roman glass producers with overlapping elemental and isotopic characteristics in such areas cannot be definitively confirmed or excluded. It is clear our large scale geological prospecting needs a detailed archaeological follow up, if glass factories are to be discovered or confirmed.

From our analyses (Fig. 7.1), it is clear glass factories in the eastern Mediterranean and possibly Italy were active from the onset of natron glass making, but likely other producers in the western Mediterranean or north Africa were also active. The earliest more negative Nd signatures observed in this study can be found from the 5th century BC onwards, but most glass seems to have an eastern Mediterranean primary origin, though a group of strong coloured (mainly black) glass with less negative ε_{Nd} and low Al_2O_3 content originated from Carthage between the 4th to 5th century BC and could have an Italian origin of the base glass.

From imperial – early Roman times onwards, throughout the 1st century AD to the first half of the 5th century AD, the origin of primary natron glass lies in the western as well as in the eastern Mediterranean and Italy. Apparently, investments were made in several glass making units all over the Empire. It is tempting to link this development in the glass industry to the end of the Roman Republic and the beginning of the Roman Empire, when Augustus introduced a series of reforms leading to a period of flourishing trade between Rome and her many and varied provinces (Lewis and Reinhold, 1966; Robinson, 1978; West, 1932), next to the impetus of the invention of glass blowing, making glass a ubiquitous utilitarian material. Most glass has a signature typical for a Syro-Palestinian or (possibly) an Italian provenance. Clear western Mediterranean or north African signatures are a minor part of the dataset. Almost all of the 'Italian' glass is colourless

(Fig. 7.2) (mostly Sb decoloured), mostly dated between the 1st and the 3rd century AD, and has a low Pb content, showing no indication of recycling or mixing. Rosenow and Rehren (2014) suggested an Egyptian primary origin of such Roman Sb decoloured glass. An additional group of Sb decoloured glass of 'Italian' signature has an extremely high Pb content of several thousand ppm, likely related to the antimony mineral used in its manufacture. The glass with a signature typical for a Syro-Palestinian origin comprises all glass types and colours.

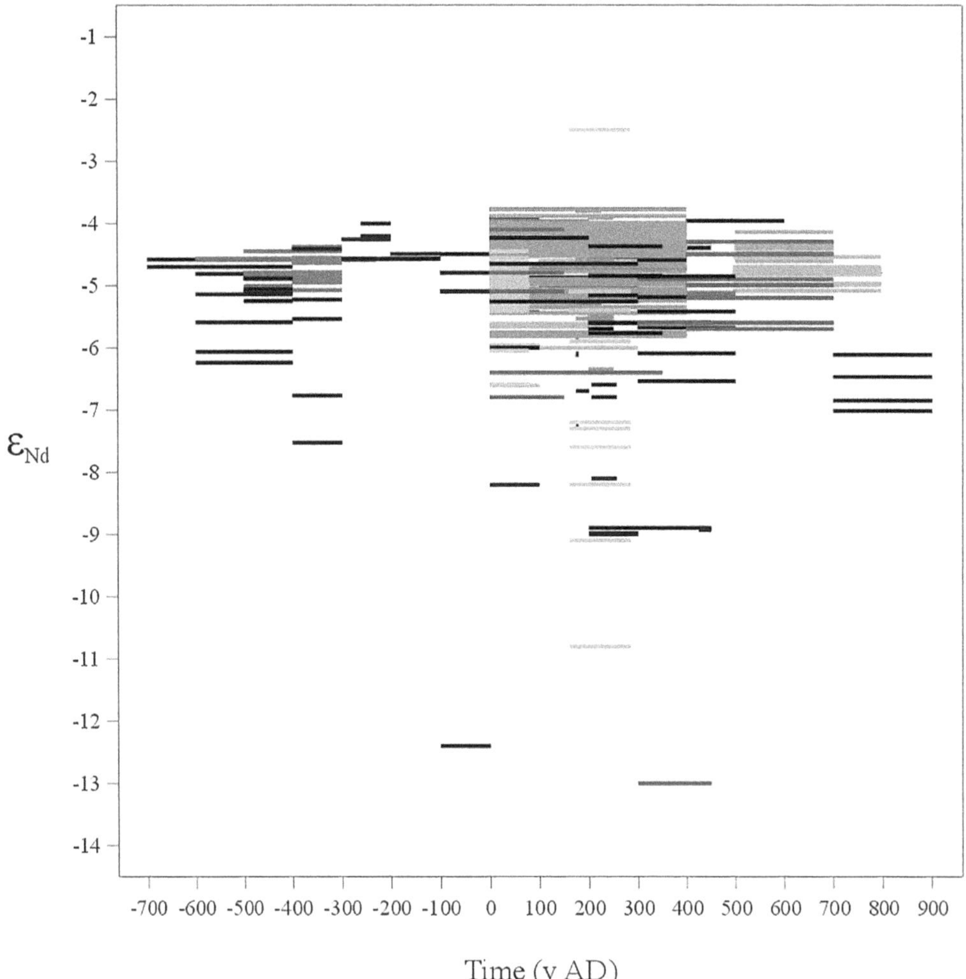

Figure 7.1:
Graphical representation of the occurrence of ε_{Nd} signatures through time for all glass analysed. Sample chronology was determined by stratigraphical association.

The presumed Italian signature of glass slowly dies out towards the end of the 5th century AD. It is tempting to relate this process to the fall of the western Roman empire (Fig. 7.1). By the 3rd century AD, the picture of peaceful commerce throughout the Roman world changes, as foreign parties exert steadily increasing pressure on the frontier regions and internal economic troubles lead to an almost empire-wide crisis (Lewis and Reinhold, 1966). Devaluation of coinage and an increase of inflation led to goods only being produced and sold locally; mass production for export was no longer possible (Cameron, 1993; Lewis and Reinhold, 1966; Robinson, 1978; Vlachou et al., 2002). A series of reforms were attempted, but the huge government expenditure for war, maintenance of the army, court and bureaucracy, and distrust for the newly introduced currency brought on a new wave of inflation (Lewis and Reinhold, 1966). In 301 AD, Diocletian issued his edict on maximum prices. By the end of the 3rd to the start of the 4th century AD, the empire is split into East and West (Lewis and Reinhold, 1966; Robinson, 1978). The turmoil of the preceding centuries, natural disasters, political upheaval, economic distress, have all been cited as potential reasons for the fall of the empire (Cameron, 1993; Lewis and Reinhold, 1966; Robinson, 1978). As the mass production and export of goods reduces or ceases in the course of the 3rd to 4th century AD, the market and transport mechanisms for glass factories around the empire would have been dissolved. The chemical signature of the western Mediterranean factories would slowly die out with recycling of old glass and the input of new raw glass from the remaining factories in the east. In late Roman to early Byzantine/Islamic times, from the 5th century AD onwards, natron glass making falls back exclusively on the glass producing sites in the eastern Mediterranean, both in Syro-Palestine and Egypt (with HIMT and Egypt II glass analysed here).

In terms of the flux source, all ancient natron glass analysed here is very homogenous in B isotopic composition. The fact that all natron sources from north Africa analysed so far are very similar (to identical) in isotopic composition, and consistent with ancient natron glass, makes placing the source of all flux for natron glass making in this area very tempting. The recent discovery of natron deposits in Fezzan shows that there may be many such sources yet undiscovered in this part of the Roman world. The occurrence and mining of natron throughout north Africa would make the supply of this raw material across the Mediterranean to primary factories in Italy of the western provinces much easier, even with overland transport. With the results obtained so far, the Wadi Natrun (and by extension North Africa) as a source of natron for Roman glass making is still the most probable.

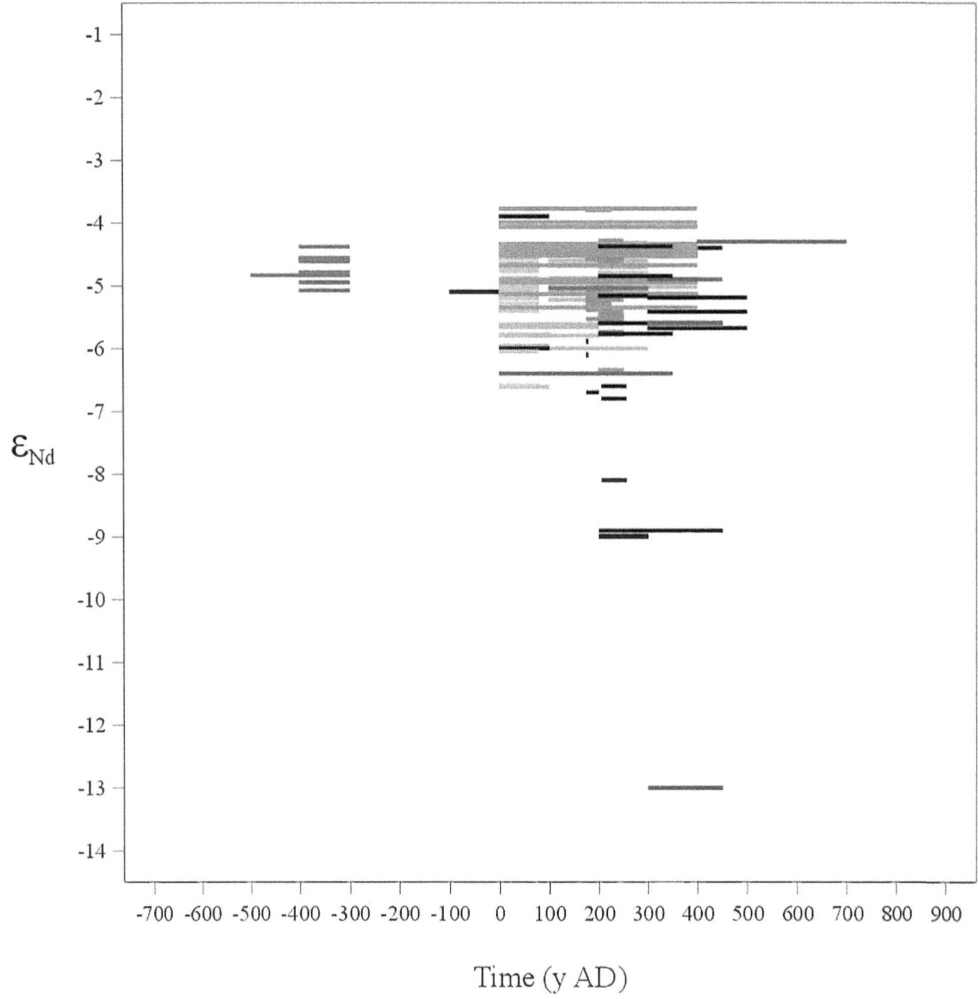

Figure 7.2:
Graphical representation of the occurrence of ε_{Nd} signatures through time for the colourless glass analysed. Sample chronology was determined by stratigraphical association.

Trace elements that proved to be the most diagnostic for provenance are Ti, Cr, Sr, Zr and Ba. Apart from Ba (and possibly Sr), which is often associated with Mn decolourants, these elements are seldom influenced by the addition of colouring agents or recycling, and provide direct information about the nature of the silica source used. Moreover, lightly elevated concentrations in glass of trace elements commonly associated with colouring agents, such as Mn, Co, Ni, Cu, Zn, Sb and Pb, are often interpreted as the result of recycling of glass cullet.

Almost a quarter of the dataset investigated appears to be recycled glass. The analysis of these elements in suitable glassmaking sands provides a good idea of the background levels that can be attributed to impurities in the source of silica. The current dataset suggests that for the two most commonly used decolourisers, MnO and Sb, these background levels are 0.1% and 30 ppm respectively. The presence of higher amounts of these elements in Roman glass would indicate their deliberate or accidental (due to recycling of cullet) addition. The definition of a high Pb Sb-decoloured glass group is very interesting in this respect, providing a new supplementary explanation for elevated Pb contents in recycled glass.

It is clear a better dating resolution of the glass samples analysed, can constrain better in time and space production units of natron glass. Also, not all issues in the technology of glass making have been unraveled, e.g. how alumina contents in raw glass are affected by the furnace material (it can be observed raw glass has higher Al_2O_3 contents than the sand used) or how trace elements are affected by recycling. The discovery of a phasing in glass making in the Hellenistic-Roman world, however, adds a new chapter to the history of glass and our knowledge of the archaeological record, now to be integrated in further economic studies of the Roman world.

Sampling locations, elemental compositions of the analysed sand samples as determined by ICP–OES analysis, LOI results, results of the Sr and Nd isotopic analysis and trace element analysis results

Sample	Beach location	Latitude (°N)	Longitude (°E)	SiO$_2$	Al$_2$O$_3$
				(Wt.%)	(Wt.%)
				ICP-OES	ICP-OES
Spain					
SP10DB46	Isla Canela	N37°10'34.08"	W007°21'15.58"	87,99	3,62
SP10DB45	Mazagón	N37°07'44.86"	W006°49'29.42"	90,73	1,85
SP10DB44	Ruinas Torre Vigia	N37°00'14.68"	W006°33'54.74"	93,84	1,28
SP10DB43	Sanlúcar de Barrameda	N36°46'48.43"	W006°21'59.54"	89,16	2,19
SP10DB42	El Puerto de Santa Maria	N36°34'40.86"	W006°13'38.77"	73,53	1,18
SP10DB41	Conil de la Frontera	N36°16'14.39"	W006°05'28.56"	55,47	1,05
SP10DB40	Barbate	N36°10'44.02"	W005°54'13.31"	57,33	0,28
SP10DB39	El Lentiscal	N36°05'19.16"	W005°46'52.05"	57,66	0,26
SP10DB38	Tarifa	N36°02'45.83"	W005°38'26.80"	73,55	0,36
SP10DB37	Carteya-Guadarranque (San Roque)	N36°10'51.65"	W005°24'45.42"	89,75	1,57
SP10DB36	Sotogrande	N36°16'44.21"	W005°16'49.73"	74,81	8,60
SP10DB35	San Pedro de Alcántara	N36°28'30.59"	W004°58'55.83"	65,33	7,66
SP10DB34	Fuengirola	N36°31'30.94"	W004°37'37.23"	66,23	5,46
SP10DB33	Guadalmar	N36°39'27.87"	W004°27'53.67"	64,43	6,99
SP10DB32	Málaga	N36°43'11.69"	W004°24'14.53"	69,41	6,43
SP10DB31	Rincón de la Victoria	N36°42'51.19"	W004°16'57.83"	66,78	11,83
SP10DB30	Punta de Torrox	N36°43'38.63"	W003°57'35.17"	58,21	19,23
SP10DB29	Salobreña	N36°43'54.59"	W003°35'16.38"	68,42	6,21
SP10DB28	Adra	N36°44'38.17"	W003°00'45.75"	82,64	5,10
SP10DB27	Almerimar	N36°42'07.46"	W002°48'03.33"	77,23	3,62
SP10DB26	Almeria	N36°49'03.19"	W002°26'15.28"	78,65	4,10
SP10DB25	El Romeral	N36°44'35.71"	W002°07'20.50"	66,94	2,78
SP10DB24	Las Negras	N36°52'47.85"	W002°00'13.85"	52,82	9,49
SP10DB23	Urbanización El Palmeral	N37°09'06.29"	W001°49'25.86"	65,89	5,93
SP10DB22	Las Marinas de Vera (Garrucha)	N37°11'50.56"	W001°48'44.82"	79,45	3,76
SP10DB21	Palomareas Bajo	N37°13'58.80"	W001°47'42.83"	66,99	6,87
SP10DB20	El Rubial	N37°24'02.10"	W001°35'33.26"	89,39	2,72
SP10DB19	Bolnuevo (Playasol)	N37°33'46.63"	W001°18'48.88"	66,15	5,96
SP10DB18	El Portús	N37°35'03.14"	W001°04'24.03"	25,61	1,47
SP08DB01	Cartagena	N37°35'	W000°58'	33,73	2,15
SP10DB17	Mil Palmeras	N37°52'54.13"	W000°45'13.59"	61,07	0,77

Fe$_2$O$_3$(t)	MgO	MnO	CaO	Na$_2$O	K$_2$O	TiO$_2$	P$_2$O$_5$	LOI	Total Oxides
(Wt.%)	(Wt.%)	(Wt.%)	(Wt.%)	(Wt.%)	(Wt.%)	(Wt.%)	(Wt.%)	(Wt.%)	
ICP-OES	ICP-OES	ICP-OES	ICP-OES	ICP-OES	ICP-OES	ICP-OES	ICP-OES		
1,37	0,37	0,04	1,57	0,63	1,00	0,17	0,04	1,50	98,28
1,01	0,17	0,01	2,07	0,31	0,67	0,09	0,04	1,94	98,89
0,47	0,07	0,01	0,76	0,19	0,45	0,05	0,02	0,82	97,96
0,49	0,15	0,01	2,38	0,58	0,93	0,06	0,03	3,29	99,27
0,75	0,65	0,04	12,02	0,24	0,45	0,06	0,04	10,14	99,10
0,47	0,52	0,02	21,43	0,40	0,38	0,09	0,06	18,34	98,24
0,36	0,30	0,02	22,07	0,20	0,09	0,01	0,05	11,56	92,28
0,37	0,30	0,02	21,97	0,20	0,09	0,01	0,04	18,27	99,19
0,59	0,22	0,01	13,56	0,13	0,12	0,02	0,03	10,69	99,28
0,82	0,79	0,02	2,52	0,31	0,21	0,08	0,03	2,48	98,58
4,42	3,33	0,15	2,64	0,55	0,74	0,27	0,05	2,89	98,45
3,55	7,51	0,05	5,83	0,62	1,05	0,27	0,06	7,57	99,50
3,28	6,64	0,05	7,55	0,65	0,76	0,21	0,05	8,30	99,17
4,05	4,73	0,09	7,08	0,60	1,12	0,34	0,07	9,13	98,62
3,68	3,25	0,07	5,90	0,59	1,01	0,28	0,09	7,18	97,89
5,44	1,36	0,13	4,46	0,66	1,35	0,48	0,10	5,80	98,39
4,52	2,67	0,11	5,19	0,42	0,60	0,48	0,07	7,01	98,49
3,01	3,50	0,06	6,67	0,75	0,98	0,29	0,08	9,42	99,38
2,78	0,79	0,03	2,16	0,70	0,75	0,28	0,08	3,05	98,35
2,25	1,59	0,06	5,88	0,53	0,58	0,18	0,05	6,90	98,86
2,46	1,28	0,04	4,49	0,55	0,68	0,34	0,06	5,74	98,40
1,90	1,69	0,04	12,63	0,45	0,47	0,67	0,05	11,40	99,02
7,09	5,99	0,13	11,72	1,47	1,69	0,58	0,07	6,91	97,96
7,06	2,36	0,28	8,08	0,54	0,47	0,30	0,06	7,78	98,75
1,88	1,06	0,04	4,77	0,57	0,59	0,19	0,05	5,44	97,79
6,56	2,10	0,23	7,41	0,71	0,59	0,36	0,07	7,41	99,31
1,34	0,52	0,02	1,21	0,34	0,50	0,15	0,05	1,91	98,15
4,75	2,47	0,19	8,54	0,80	0,91	0,27	0,08	8,27	98,39
1,63	9,18	0,05	27,96	0,24	0,32	0,11	0,05	31,80	98,43
1,98	5,99	0,05	28,25	0,40	0,41	0,20	0,04	27,03	100,22
0,22	1,50	0,01	18,58	0,09	0,40	0,04	0,02	16,51	99,22

Sample	Beach location	Latitude (°N)	Longitude (°E)	SiO$_2$	Al$_2$O$_3$
				(Wt.%)	(Wt.%)
				ICP-OES	ICP-OES
SP10DB16	Guardamar del Segura	N38°05'55.00"	W000°38'42.46"	44,76	1,39
SP10DB15	Urbanova (Alicante)	N38°16'59.85"	W000°31'12.75"	22,75	0,38
SP10DB14	Villajoyosa	N38°30'23.71"	W000°13'38.11"	62,76	0,70
SP10DB13	Calpe	N38°38'30.85"	E000°03'15.13"	55,98	1,31
SP10DB12	Dénia	N38°51'09.47"	E000°05'45.90"	18,56	0,62
SP10DB11	Platja de Daimús	N38°58'45.16"	W000°08'37.81"	59,40	0,93
SP10DB10	Cullera	N39°09'23.94"	W000°14'24.51"	62,06	1,95
SP10DB09	Valencia	N39°24'27.55"	W000°19'54.16"	75,12	0,99
SP10DB08	Sagunto	N39°39'36.15"	W000°12'34.58"	63,88	1,97
SP10DB07	Burriana	N39°51'55.76"	W000°03'53.92"	63,92	1,76
SP10DB06	Castellón de la Plana	N39°59'44.86"	E000°01'43.78"	59,35	3,17
SP08DB02	Benicassim (Castellon)	N40°02'	E000°03'	60,65	3,12
SP10DB05	Torreblanca	N40°14'02.18"	E000°16'21.75"	60,31	3,97
SP10DB04	Benicarló	N40°25'09.84"	E000°26'13.06"	47,51	2,14
SP08FH05	Platja del Trabucador	N40°38'04.7"	E000°44'57.6"	39,80	3,80
SP08FH06	Riumar	N40°43'49.9"	E000°50'30.0"	62,53	3,63
SP08FH07	Marquesa	N40°45'42.9"	E000°47'53.1"	51,35	3,52
SP08FH08	Cambrils	N41°03'57.6"	E001°04'06.3"	76,43	7,85
SP10DB47	Coma-Ruga	N41°11'00.68"	E001°32'46.71"	73,47	5,01
SP08FH04	Castelldefels	N41°15'51.7"	E001°57'02.0"	71,80	4,98
SP08FH03	Vilassar de Mar	N41°30'46.6"	E002°24'46.3"	80,05	11,40
SP08FH01	Canet de Mar	N41°35'10.5"	E002°34'56.6"	83,51	9,00
SP08FH02	Malgrat de Mar	N41°38'21.8"	E002°44'17.6"	83,10	9,08
SP08FH12	Sa Conca	N41°47'52.05"	E003°03'38.31"	87,36	6,72
SP08FH11	Platja d'Aro	N41°49'02.08"	E003°04'11.10"	84,24	6,41
SP08FH10	Sant Pere Pescador	N42°11'21.43"	E003°06'38.18"	83,29	8,92
SP08FH09	Canyelles Petites	N42°14'27.4"	E003°12'26.2"	80,31	9,75
SP09DB03	Colera	N42°24'22.75"	E003°09'20.25"	71,46	10,99
SP10DB48	Cavalleria, Menorca	N40°03'34.33"	E004°04'35.42"		
SP10DB49	Cala Tirant, Furnelles, Menorca	N40°02'38.84"	E004°06'16.73"		
France					
FR09DB02	Biarritz, St-Jean L(h)uzz	N43°28'19"	W001°34'12"	82,79	0,95
FR09DB03	Argelès-Plage, Argelès-sur-Mer	N42°34'38.33"	E003°02'43.94"	80,12	10,10
FR09DB04	Le Barcarès	N42°46'57.21"	E003°02'20.30"	82,43	8,97
FR09DB05	Port-la-Nouvelle	N43°01'46.30"	E003°04'03.88"	54,62	5,02
FR09DB06	Les Cabanes de Fleury	N43°12'24.34"	E003°13'44.03"	68,99	6,94
FR09DB07	Sérignan-Plage (Valras-Plage)	N43°14'56.28"	E003°18'07.22"	62,76	7,05
FR09DB08	Le Grau-d'Agde	N43°16'57.91"	E003°26'59.95"	65,58	4,84
FR09DB09	Frontignan Plage	N43°25'39.52"	E003°45'37.49"	53,71	4,01
FR09DB10	Saintes-Maries-de-la-Mer	N43°26'57.43"	E004°24'57.79"	74,92	5,44
FR09DB11	Salin-de-Giraud (Port-Saint-Louis-du-Rhône)	N43°20'41.99"	E004°47'47.02"	52,75	6,06
FR09DB12	Sainte-Croix, Martigues	N43°19'53.99"	E005°04'15.63"	12,02	0,42
FR09DB13	L'Estaque, Plage des Corbières	N43°21'27.36"	E005°17'26.53"	88,37	1,79

$Fe_2O_3(t)$	MgO	MnO	CaO	Na_2O	K_2O	TiO_2	P_2O_5	LOI	Total Oxides
(Wt.%)	(Wt.%)	(Wt.%)	(Wt.%)	(Wt.%)	(Wt.%)	(Wt.%)	(Wt.%)	(Wt.%)	
ICP-OES	ICP-OES	ICP-OES	ICP-OES	ICP-OES	ICP-OES	ICP-OES	ICP-OES		
0,68	4,27	0,02	23,69	0,15	0,64	0,08	0,04	23,31	99,03
0,38	1,08	0,01	39,34	0,15	0,13	0,02	0,04	33,78	98,05
0,29	0,51	0,01	19,08	0,08	0,25	0,04	0,03	15,81	99,55
0,73	0,91	0,02	21,94	0,15	0,32	0,10	0,03	18,08	99,59
0,60	2,73	0,02	38,90	0,25	0,29	0,02	0,05	35,50	97,55
0,55	1,37	0,02	19,30	0,15	0,42	0,04	0,03	17,28	99,49
0,73	0,74	0,02	17,67	0,19	0,80	0,08	0,04	14,79	99,07
0,42	0,40	0,01	12,53	0,12	0,46	0,06	0,03	10,14	100,28
0,94	0,68	0,02	16,77	0,23	0,51	0,09	0,04	13,62	98,75
0,94	0,54	0,02	17,30	0,21	0,52	0,11	0,04	14,07	99,43
0,96	0,73	0,02	19,00	0,40	0,75	0,13	0,04	15,38	99,96
1,46	0,87	0,03	19,03	0,37	0,65	0,39	0,05	16,56	103,17
1,95	1,01	0,04	17,77	0,37	0,63	0,47	0,06	13,76	100,34
1,06	0,59	0,01	25,93	0,32	0,41	0,06	0,04	21,16	99,22
2,74	1,57	0,05	28,03	0,33	0,59	0,27	0,07	22,32	99,58
1,52	0,69	0,03	16,16	0,60	1,02	0,14	0,05	13,63	100,00
2,01	1,08	0,04	22,13	0,65	0,78	0,24	0,06	18,87	100,74
1,33	0,78	0,02	4,91	1,28	2,89	0,16	0,05	4,36	100,06
0,85	0,74	0,01	8,25	0,96	1,52	0,06	0,05	7,13	98,04
1,57	1,21	0,02	8,92	0,88	1,62	0,13	0,05	8,64	99,82
0,80	0,23	0,01	1,74	2,58	3,13	0,05	0,03	1,06	101,07
0,80	0,16	0,02	0,78	2,11	2,76	0,04	0,03	0,55	99,76
0,81	0,18	0,01	0,56	1,93	3,16	0,06	0,03	0,60	99,53
0,56	0,06	0,01	0,33	1,37	2,89	0,02	0,03	0,33	99,69
0,58	0,07	0,01	0,33	1,36	2,61	0,03	0,03	0,35	96,03
0,56	0,14	0,01	0,68	2,00	3,38	0,03	0,03	0,69	99,73
1,41	0,32	0,02	1,62	3,16	2,35	0,10	0,03	1,20	100,28
4,86	1,71	0,05	2,44	1,57	2,25	0,51	0,14	3,97	99,93
0,50	0,23	0,01	8,56	0,29	0,24	0,06	0,03	7,49	101,15
2,21	0,73	0,03	0,52	1,52	3,38	0,25	0,08	1,36	100,30
1,90	0,52	0,03	0,97	1,35	3,00	0,16	0,05	1,23	100,61
1,44	1,68	0,06	19,27	1,22	1,31	0,27	0,08	16,39	101,37
2,09	1,74	0,04	8,56	0,99	2,38	0,25	0,08	8,94	101,01
2,03	2,49	0,04	10,82	1,03	2,51	0,26	0,09	11,13	100,20
1,19	0,85	0,04	14,64	1,27	1,32	0,28	0,09	11,67	101,77
0,88	0,58	0,03	22,21	1,14	1,34	0,14	0,06	17,80	101,89
1,08	0,56	0,03	8,81	1,21	1,56	0,22	0,07	7,31	101,21
4,37	2,42	0,13	18,52	0,90	0,98	1,77	0,36	13,35	101,59
0,30	1,51	0,02	47,85	0,48	0,12	0,04	0,22	38,46	101,43
0,16	0,30	0,00	3,74	0,73	1,28	0,04	0,02	4,56	100,99

Sample	Beach location	Latitude (°N)	Longitude (°E)	SiO$_2$	Al$_2$O$_3$
				(Wt.%)	(Wt.%)
				ICP-OES	ICP-OES
FR09DB14	La Ciotat	N43°11'06.47"	E005°37'11.86"	30,30	0,58
FR09DB16	Les Bormettes, La Londe-les-Maures	N43°07'16.69"	E006°15'38.86"	90,02	4,21
FR09DB17	Cavalaire-sur-Mer	N43°10'57.28"	E006°32'28.29"	85,67	7,75
FR09DB18	Saint-Aygulf	N43°24'31.78"	E006°44'06.80"	70,28	6,47
FR09DB19	Cannes	N43°32'55.72"	E007°00'13.63"	77,51	5,36
FR09DB20	Saint-Laurent-du-Var	N43°39'26.11"	E007°11'45.98"	50,86	5,62
FR09DB01	Menton	N43°46'40.18"	E007°30'29.67"	58,85	5,12
Italy					
IT09DB12	Pigna, Andora SV	N43°56'49.79"	E008°08'27.84"	67,95	2,81
IT09DB13	Finale Pia, Finale Ligure	N44°10'18.48"	E008°21'30.08"	77,15	4,19
IT09DB14	Albissola Marina, Albisola Superiore	N44°19'38.02"	E008°30'17.23"	79,95	8,53
IT09DB15	Varazze	N44°21'27.59"	E008°34'20.90"	75,47	7,72
IT09DB16a	Voltri	N44°25'38.18"	E008°44'44.51"	51,57	8,61
IT09DB16b	Voltri	N44°25'38.18"	E008°44'44.51"	51,55	6,49
IT09DB17	Sestri Levante	N44°16'21.51"	E009°23'35.56"	64,15	7,44
IT09DB18	Levanto	N44°10'17.18"	E009°36'23.54"	51,64	11,36
IT09DB19	Fiumaretta di Ameglia	N44°02'56.69"	E009°59'27.63"	65,48	8,14
IT09DB09	Torre Del Lago Puccini, Viareggio	N43°49'28.23"	E010°15'15.36"	70,53	6,54
IT10DB30	Migliarino, Vecchiano	N43°47'33.04"	E010°15'55.95"	73,53	6,78
IT10DB31	Marina di Pisa	N43°40'00.06"	E010°16'30.89"	71,32	6,55
IT10DB32	Castiglioncello	N43°24'23.18"	E010°24'27.77"	29,49	5,78
IT10DB33	Marina di Cecina	N43°17'47.37"	E010°29'41.12"	51,49	5,97
IT08DB05	Elba	N42°47'34.75"	E010°14'57.82"	72,83	9,92
IT10DB34	Torre del Sale, Piombino	N42°57'14.50"	E010°36'00.71"	79,29	2,83
IT08DB01	Cala Violina	N42°50'19.53"	E010°46'29.46"	92,57	3,12
IT10DB35	Castiglione della Pescaia	N42°45'43.06"	E010°52'36.13"	55,18	4,63
IT10DB36	Principina a Mare, Grosseto	N42°41'21.78"	E010°59'50.59"	49,48	4,48
IT10DB37	Albinia	N42°30'02.56"	E011°11'32.49"	44,51	4,51
IT10DB38	Poggio Pertuso, Monte Argentario	N42°24'30.22"	E011°12'33.40"	53,43	8,85
IT10DB39	Montalto Marina	N42°19'41.71"	E011°34'29.44"	54,67	12,58
IT10DB40	Voltone, Tarquinia	N42°14'22.33"	E011°41'25.08"	51,46	9,85
IT10DB41	Ladispoli	N41°57'00.78"	E012°04'10.54"	52,59	6,49
IT10DB42	Lido di Ostia	N41°44'03.23"	E012°15'41.94"	49,15	5,76
IT08DB06	Ostia	N41°43'41.45"	E012°16'40.78"	49,85	5,79
IT10DB43	Anzio	N41°27'09.47"	E012°38'22.08"	48,34	5,02
IT10DB44	Terracina	N41°16'50.95"	E013°11'47.24"	60,83	2,08
IT09DB22	Gaeta (BRILL 4556; Degryse and Schneider, 2008)	N41°12'47"	E013°32'33"	nd.	nd.
IT10DB45	Marina di Mintuno	N41°13'28.28"	E013°45'33.66"	50,85	5,44
IT09DB23	Volturno, coastal strip of Mondragone (MN1; Silvestri et al., 2006)	N41°07'32.85"	E013°51'55.31"	51,67	5,27

$Fe_2O_3(t)$	MgO	MnO	CaO	Na_2O	K_2O	TiO_2	P_2O_5	LOI	Total Oxides
(Wt.%)	(Wt.%)	(Wt.%)	(Wt.%)	(Wt.%)	(Wt.%)	(Wt.%)	(Wt.%)	(Wt.%)	
ICP-OES	ICP-OES	ICP-OES	ICP-OES	ICP-OES	ICP-OES	ICP-OES	ICP-OES		
0,18	0,63	0,00	38,52	0,17	0,28	0,03	0,01	30,44	101,14
1,65	0,27	0,01	0,42	1,11	0,81	0,29	0,04	1,62	100,44
1,51	0,46	0,04	1,05	2,30	0,78	0,43	0,03	0,58	100,60
0,84	0,76	0,02	10,02	1,37	2,23	0,12	0,04	9,07	101,23
0,56	0,55	0,01	7,05	1,05	2,09	0,08	0,03	6,28	100,57
1,59	0,92	0,03	21,41	1,16	1,54	0,18	0,12	17,21	100,64
0,55	1,15	0,01	17,27	1,07	2,15	0,08	0,08	14,43	100,77
1,05	0,39	0,04	14,90	0,72	0,75	0,10	0,03	12,10	100,85
1,30	2,13	0,03	6,60	1,02	0,97	0,15	0,04	6,95	100,53
2,96	1,90	0,05	2,21	2,05	1,62	0,62	0,11	2,21	102,22
3,20	3,83	0,06	3,35	1,81	1,05	0,48	0,06	3,28	100,28
12,31	12,69	0,15	4,44	1,22	0,91	0,92	0,06	6,83	99,69
7,12	20,22	0,13	3,75	0,85	0,70	0,46	0,04	9,09	100,42
6,10	11,32	0,07	2,07	1,89	0,85	0,40	0,04	6,22	100,54
6,18	13,83	0,11	6,43	2,09	0,66	0,36	0,03	7,57	100,25
3,01	3,40	0,06	8,54	1,69	1,74	0,21	0,05	8,49	100,82
2,21	0,99	0,09	8,46	1,60	1,54	0,36	0,04	7,37	99,73
1,90	0,88	0,08	6,28	1,76	1,85	0,15	0,06	5,98	99,24
1,99	0,94	0,08	8,31	1,72	1,63	0,19	0,05	7,34	100,12
3,12	6,89	0,16	26,63	1,54	0,24	0,18	0,03	24,36	98,44
5,17	10,26	0,13	11,76	1,18	0,51	0,27	0,05	12,43	99,22
3,25	1,74	0,05	4,72	1,94	2,92	0,40	0,06	3,03	100,86
0,99	0,34	0,08	7,97	0,70	0,96	0,08	0,02	6,44	99,69
0,96	0,23	0,02	0,76	0,55	0,83	0,34	0,01	0,50	99,88
2,96	1,34	0,22	17,98	0,97	0,96	0,36	0,05	15,22	99,89
3,97	1,39	0,16	18,33	0,71	0,83	0,23	0,05	16,20	95,82
3,86	1,17	0,25	23,35	0,58	0,92	0,15	0,05	19,89	99,23
2,96	1,49	0,18	14,38	1,19	3,56	0,23	0,06	12,47	98,81
4,29	3,45	0,13	11,85	1,43	5,05	0,41	0,11	5,89	99,85
4,86	5,96	0,12	16,32	0,96	3,73	0,47	0,11	5,38	99,21
8,66	8,66	0,16	19,75	0,69	0,86	0,96	0,12	1,98	100,92
7,83	10,71	0,15	22,18	0,47	0,66	0,78	0,13	3,19	101,02
8,80	7,89	0,18	20,98	0,58	0,73	0,98	0,12	6,26	102,15
1,96	1,31	0,09	13,23	1,10	1,35	0,23	0,09	11,93	84,65
1,08	1,39	0,04	16,98	0,48	1,08	0,05	0,04	15,11	99,17
nd.	nd.	nd.	nd.	nd.	nd.	nd.	nd.	nd.	nd.
3,64	4,57	0,09	20,41	1,02	1,63	0,40	0,09	12,88	101,03
3,06	3,80	0,11	20,84	0,73	1,37	0,43	0,09	13,39	100,77

Sample	Beach location	Latitude (°N)	Longitude (°E)	SiO$_2$	Al$_2$O$_3$
				(Wt.%)	(Wt.%)
				ICP-OES	ICP-OES
IT09DB21	Castel Volturno, river Volturno mouth (BRILL 4554; Degryse and Schneider, 2008)	N41°01'15"	E013°55'50"	nd.	nd.
IT09DB25	Volturno, Pineta Grande (LD2)	N41°00'36.90"	E013°56'28.71"	60,71	6,94
IT09DB26	Volturno, Marina di Varcaturo/ Liternum (LD3)	N40°53'46.06"	E014°02'00.87"	59,78	6,58
IT09DB27	Volturno, Licola mare (LD4)	N40°52'47.43"	E014°02'25.59"	58,67	7,97
IT09DB20	Licola Mare (BRILL 4553; Degryse and Schneider, 2008)	N40°52'5"	E014°02'37"	nd.	nd.
IT09DB28	Volturno, Cuma (LD5)	N40°51'18.75"	E014°02'48.46"	56,68	6,33
IT09DB24	Volturno, base of promontory of Capo Miseno (CM1; Silvestri et al., 2006)	N40°47'06.53"	E014°05'02.94"	58,60	10,59
IT08DB02	Amalfi	N40°37'07.16"	E014°34'38.54"	28,52	7,10
IT10DB46	Foce del Sele	N40°28'39.31"	E014°56'44.53"	39,29	3,15
IT10DB47	Foce, Marina di Casal Velino	N40°09'41.95"	E015°08'41.66"	79,96	6,27
IT10DB48	Sapri	N40°04'14.54"	E015°37'52.88"	40,17	3,23
IT10DB49	Santa Maria del Cedro	N39°46'07.40"	E015°47'49.21"	33,14	1,20
IT10DB50	Paola	N39°22'00.36"	E016°01'38.86"	59,05	14,62
IT10DB51	Villaggio del Golfo	N39°01'09.85"	E016°06'30.01"	70,09	12,86
IT10DB52	Pizzo	N38°46'20.22"	E016°11'50.19"	79,44	10,15
IT10DB53	San Ferdinando	N38°29'54.04"	E015°55'03.13"	77,89	10,76
IT10DB54	Villafranca Tirrena	N38°15'15.19"	E015°27'14.84"	68,48	9,28
IT10DB55	Falcone	N38°07'23.05"	E015°06'15.53"	76,52	9,23
IT10DB56	Scafa, Brolo	N38°09'19.65"	E014°47'57.53"	77,14	8,50
IT10DB57	Sant'Agata di Militello	N38°05'00.42"	E014°39'07.72"	63,59	3,53
IT10DB58	Finale, Pollina	N38°01'10.90"	E014°10'57.04"	68,78	4,31
IT10DB59	San Nicola l'Arena	N38°00'35.31"	E013°37'26.32"	76,20	0,69
IT10DB60	Ficarazzi	N38°05'40.45"	E013°26'42.19"	38,96	0,64
IT10DB61	Capaci	N38°10'56.33"	E013°13'56.31"	21,30	0,17
IT10DB62	Castellammare del Golfo	N38°01'28.21"	E012°54'20.29"	87,45	0,51
IT10DB63	Marausa, Trapani	N37°56'05.62"	E012°28'53.98"	85,89	0,38
IT10DB64	Marinella	N37°34'56.66"	E012°51'47.64"	44,57	0,51
IT10DB65	Ribera	N37°27'09.18"	E013°13'25.80"	35,76	0,21
IT10DB66	Montallegro	N37°22'51.29"	E013°18'23.83"	46,89	0,73
IT08DB08	Torre Salsa	N37°22'12.78"	E013°19'01.45"	56,67	0,95
IT10DB67	Licata	N37°06'27.77"	E013°58'15.51"	55,60	1,33
IT10DB68	Marina di'Acate	N36°58'58.28"	E014°21'31.86"	61,53	0,97
IT10DB69	Plaia Grande	N36°46'10.88"	E014°36'34.84"	57,75	1,11
IT10DB70	Lido di Noto	N36°50'21.52"	E015°06'24.53"	1,55	0,40
IT10DB71	Catania	N37°27'08.69"	E015°05'11.40"	68,98	1,31
IT10DB72	San Marco, Calatabiano	N37°48'25.24"	E015°15'26.80"	71,28	10,49
IT10DB73	Nizza di Sicilia	N38°00'00.12"	E015°25'14.82"	65,06	12,44
IT10DB74	Pellaro	N38°01'08.82"	E015°38'06.25"	72,35	11,25

$Fe_2O_3(t)$	MgO	MnO	CaO	Na_2O	K_2O	TiO_2	P_2O_5	LOI	Total Oxides
(Wt.%)	(Wt.%)	(Wt.%)	(Wt.%)	(Wt.%)	(Wt.%)	(Wt.%)	(Wt.%)	(Wt.%)	
ICP-OES	ICP-OES	ICP-OES	ICP-OES	ICP-OES	ICP-OES	ICP-OES	ICP-OES		
nd.	nd.	nd.	nd.	nd.	nd.	nd.	nd.	nd.	nd.
2,39	2,36	0,11	15,01	1,08	2,25	0,25	0,09	9,06	100,25
2,81	1,68	0,13	15,37	1,11	2,11	0,35	0,11	10,60	100,64
1,84	1,04	0,14	15,15	1,39	2,93	0,18	0,08	11,80	101,19
nd.	nd.	nd.	nd.	nd.	nd.	nd.	nd.	nd.	nd.
1,41	1,13	0,14	17,65	1,17	2,15	0,14	0,07	13,79	100,67
2,14	1,20	0,12	12,26	1,87	3,60	0,22	0,09	9,38	100,06
4,49	7,91	0,08	27,95	0,85	2,25	0,50	0,21	19,61	99,48
2,71	5,01	0,13	24,92	0,51	0,70	0,21	0,13	20,55	97,31
1,93	0,68	0,04	2,99	1,29	2,02	0,14	0,04	3,02	98,40
2,95	1,86	0,05	25,79	0,48	0,46	0,16	0,05	22,96	98,17
1,17	11,73	0,05	21,49	0,30	0,15	0,06	0,04	29,00	98,34
6,94	3,22	0,11	5,82	2,14	1,20	0,82	0,11	4,44	98,47
4,28	1,94	0,06	1,50	1,95	2,51	0,52	0,11	2,96	98,78
1,07	0,37	0,01	1,97	2,09	2,88	0,13	0,07	0,83	99,02
1,11	0,36	0,02	1,53	2,75	2,68	0,13	0,07	1,04	98,35
1,37	0,46	0,02	1,40	2,36	2,23	0,15	0,07	1,05	86,87
0,96	0,35	0,03	2,82	2,38	2,84	0,11	0,06	2,62	97,91
2,89	0,94	0,04	1,49	1,87	2,00	0,25	0,08	2,30	97,51
3,31	1,07	0,15	12,77	0,66	0,41	0,28	0,06	11,52	97,35
5,70	1,35	0,08	7,36	0,44	0,50	0,29	0,10	8,89	97,78
1,09	0,82	0,04	8,65	0,21	0,12	0,05	0,03	8,48	96,37
1,26	4,70	0,02	26,12	0,11	0,12	0,04	0,10	25,57	97,66
0,24	3,29	0,01	35,48	0,28	0,04	0,01	0,06	33,93	94,81
0,98	0,33	0,02	4,66	0,29	0,16	0,08	0,03	4,40	98,92
0,53	0,12	0,02	5,25	0,73	0,14	0,07	0,02	5,51	98,66
1,01	0,44	0,05	26,40	0,20	0,15	0,02	0,11	22,39	95,86
0,35	5,94	0,01	25,63	0,17	0,06	0,01	0,04	27,18	95,37
1,34	0,52	0,07	24,70	0,25	0,20	0,03	0,06	21,37	96,16
1,40	0,43	0,08	22,42	0,24	0,28	0,04	0,05	18,71	101,27
1,19	0,31	0,07	13,94	0,32	0,39	0,05	0,04	13,06	86,29
1,17	0,39	0,08	15,85	0,43	0,30	0,05	0,08	14,01	94,85
1,32	0,53	0,10	20,31	0,29	0,32	0,07	0,06	16,27	98,14
0,42	1,42	0,02	49,71	0,10	0,07	0,03	0,06	42,40	96,19
3,65	1,51	0,10	9,49	0,25	0,16	0,31	0,05	7,91	93,73
5,77	1,48	0,07	1,02	1,32	1,94	0,39	0,11	3,76	97,64
5,10	1,81	0,11	2,12	2,18	1,82	0,65	0,13	2,84	94,25
4,60	0,94	0,32	2,01	1,86	2,13	0,37	0,14	2,02	97,98

Sample	Beach location	Latitude (°N)	Longitude (°E)	SiO_2	Al_2O_3
				(Wt.%)	(Wt.%)
				ICP-OES	ICP-OES
IT10DB75	Bova Marina	N37°55'37.60"	E015°53'24.66"	72,35	13,09
IT10DB76	San Nicola, Bovalino	N38°07'34.79"	E016°09'38.24"	71,64	12,20
IT10DB77	Marina di Caulonia	N38°20'40.60"	E016°28'25.64"	72,91	13,10
IT10DB78	Isca Marina	N38°36'23.69"	E016°33'56.88"	66,30	10,60
IT10DB79	Catanzaro Lido	N38°49'48.45"	E016°38'38.17"	73,90	13,93
IT10DB80	Steccato	N38°55'58.25"	E016°55'12.79"	76,21	11,45
IT10DB81	Tronca	N39°14'37.24"	E017°06'44.20"	76,94	10,77
IT10DB82	Mirto, Crosia	N39°37'26.14"	E016°45'25.12"	67,74	9,89
IT10DB83	Laghi di Sibari	N39°43'34.51"	E016°31'35.58"	70,29	11,97
IT10DB84	Lido di Policoro	N40°11'53.62"	E016°43'37.65"	64,31	3,20
IT10DB85	Metaponto Lido	N40°20'34.11"	E016°49'23.68"	76,85	2,69
IT09DB11	Castellaneta Marina	N40°27'45.63"	E016°56'18.41"	71,94	4,05
IT10DB86	Torre Colimena	N40°18'02.23"	E017°43'22.46"	5,96	0,32
IT10DB87	Masseria Maime	N40°33'27.00"	E018°02'36.73"	79,97	1,26
IT10DB88	Torre Canne	N40°49'57.29"	E017°28'19.07"	16,63	0,86
IT10DB89	Magherita di Savoia	N41°22'36.77"	E016°09'06.35"	62,96	4,61
IT10DB90	Siponto, Manfredonia	N41°35'32.58"	E015°53'41.45"	67,21	6,05
IT08DB07	Gargano	N41°56'11.64"	E015°56'50.84"	50,03	4,39
IT10DB91	Campomarino	N41°58'39.04"	E015°01'52.56"	66,35	4,42
IT10DB92	Pescara	N42°28'28.71"	E014°12'40.47"	51,40	4,38
IT10DB93	Sentina, San Benedetto del Tronto	N42°54'39.67"	E013°54'27.58"	56,75	3,94
IT08DB04	San Benedetto del Tronto	N42°57'57.45"	E013°52'48.24"	64,65	3,51
IT10DB94	Civitanova Marche	N43°17'44.25"	E013°44'30.80"	59,31	2,91
IT10DB95	Senigallia	N43°42'17.76"	E013°14'28.28"	54,95	4,23
IT10DB96	Gabicce Mare	N43°58'03.83"	E012°44'25.39"	37,77	3,24
IT10DB97	Cesenatico	N44°12'48.47"	E012°23'29.67"	55,75	7,09
IT08DB03	Uniti estuary, Ravenna	N44°23'38.99"	E012°18'56.81"	51,90	6,21
IT10DB98	Casalborsetti	N44°33'11.27"	E012°17'02.37"	61,32	7,01
IT09DB10	Casalborsetti	N44°33'17.84"	E012°17'00.57"	64,41	7,69
IT10DB99	Lido delle Nazioni	N44°43'54.17"	E012°14'31.82"	68,53	7,89
IT10DB100	Barricata, Bonelli	N44°50'34.06"	E012°27'36.74"	68,39	9,08
IT10DB101	Boccasette	N45°01'37.79"	E012°25'25.92"	66,36	9,04
IT10DB102	Bacucco, Chioggia	N45°10'21.51"	E012°19'30.55"	43,22	8,07
IT10DB103	Cortellazzo	N45°31'43.61"	E012°43'28.45"	7,68	0,85
IT10DB104	Bibione, Lignano Sabbiadoro	N45°37'53.58"	E013°03'23.80"	10,77	1,05
IT10DB105	Palau, NE Sardinia	N41°10'58.59"	E009°22'20.70"		
IT10DB106	Alghero, NW Sardinia	N40°34'02.30"	E008°19'05.83"		
IT10DB107	Poetto, SE Sardinia	N39°12'09.26"	E009°09'48.20"		
IT10DB108	Solanas, SE Sardinia	N39°08'04.99"	E009°25'41.71"		
IT09DB29	Rocca Ruja, Sardinia	N40°57'36"	E008°12'46"	21,12	2,25
Shell				0,68	0,04
Na_2CO_3				0,29	0,00

Fe$_2$O$_3$(t)	MgO	MnO	CaO	Na$_2$O	K$_2$O	TiO$_2$	P$_2$O$_5$	LOI	Total Oxides
(Wt.%)	(Wt.%)	(Wt.%)	(Wt.%)	(Wt.%)	(Wt.%)	(Wt.%)	(Wt.%)	(Wt.%)	
ICP-OES	ICP-OES	ICP-OES	ICP-OES	ICP-OES	ICP-OES	ICP-OES	ICP-OES		
2,94	1,05	0,06	1,50	3,29	2,10	0,36	0,11	1,17	98,02
3,78	1,31	0,06	1,11	2,31	2,31	0,48	0,11	1,62	96,94
2,10	0,70	0,03	0,89	3,02	3,07	0,21	0,07	1,36	97,45
1,63	0,52	0,03	0,83	2,37	2,65	0,17	0,06	0,98	86,15
2,37	1,00	0,03	1,98	2,93	2,73	0,31	0,08	1,10	100,37
1,14	0,34	0,02	2,20	2,37	3,06	0,11	0,09	1,39	98,39
2,00	0,62	0,03	1,65	2,06	3,00	0,23	0,08	1,56	98,93
2,94	0,97	0,04	1,05	1,85	2,21	0,28	0,10	2,06	89,13
3,19	1,28	0,06	2,19	2,29	1,88	0,51	0,08	2,29	96,01
2,66	3,43	0,08	8,62	0,54	0,55	0,12	0,04	9,21	92,75
1,46	0,59	0,09	9,37	0,64	0,89	0,11	0,05	7,73	100,47
1,00	0,35	0,09	11,59	0,87	1,58	0,06	0,02	9,32	100,87
0,26	2,45	0,03	45,79	0,32	0,11	0,02	0,04	41,34	96,62
0,54	0,31	0,02	8,95	0,34	0,69	0,04	0,03	7,48	99,63
0,41	1,74	0,03	42,64	0,55	0,36	0,04	0,11	36,33	99,70
2,83	2,42	0,14	16,69	0,70	1,30	0,31	0,16	10,04	102,16
1,31	0,55	0,13	12,17	1,37	2,57	0,10	0,08	10,10	101,64
0,97	0,63	0,19	22,88	0,92	1,88	0,07	0,03	18,18	100,19
0,86	0,37	0,10	14,60	0,99	1,88	0,05	0,05	11,34	101,02
0,94	0,94	0,06	22,28	1,03	1,60	0,07	0,06	18,07	100,82
0,98	1,17	0,04	19,82	1,01	0,96	0,14	0,06	16,10	100,95
0,58	0,42	0,03	16,54	0,90	1,29	0,09	0,03	12,54	100,58
1,06	0,30	0,04	18,74	0,72	0,94	0,09	0,04	15,07	99,21
0,83	0,38	0,08	20,65	1,06	1,67	0,06	0,05	15,98	99,93
1,58	1,95	0,12	29,13	0,77	0,72	0,19	0,12	24,67	100,26
2,49	1,68	0,12	15,42	1,50	1,17	0,44	0,17	12,46	98,28
2,19	1,53	0,11	19,64	1,38	1,28	0,29	0,07	16,24	100,83
1,83	1,08	0,09	11,01	1,63	1,64	0,26	0,10	9,62	95,59
1,83	1,26	0,09	11,54	1,82	1,84	0,25	0,08	9,75	100,56
1,72	1,26	0,07	9,87	1,88	1,86	0,18	0,08	8,13	101,47
2,60	2,27	0,07	7,76	2,06	1,77	0,28	0,09	6,97	101,33
3,20	2,41	0,09	8,10	1,83	1,54	0,41	0,13	6,77	99,89
3,29	7,53	0,07	16,81	1,18	1,52	0,54	0,10	18,01	100,33
0,64	13,98	0,03	30,48	0,18	0,18	0,08	0,03	39,64	93,78
0,71	14,08	0,02	31,47	0,22	0,20	0,07	0,03	38,28	96,91
0,60	3,62	0,02	37,22	0,80	0,57	0,15	0,05	33,07	99,48
0,01	0,06	0,00	53,77	0,72	0,03	0,00	0,02	44,51	99,84
0,00	0,01	0,00	0,00	45,74	0,00	0,00	0,01	53,95	100,00

Sample	Beach location	Latitude (°N)	Longitude (°E)	SiO$_2$	Al$_2$O$_3$
				(Wt.%)	(Wt.%)
				ICP-OES	ICP-OES
PO09DB01	Peniche	N39°21'	W009°21'	89,10	1,28
GR08DB01	Acharavi, Korfu	N39°47'54"	E019°49'01"	53,84	2,07
GR09DB02	Sporades, Alonissos, Chrissi Milia	N39°09'46"	E023°53'52"	17,30	0,64
GR09DB03	Maronia, Trakia, jun 2005	N40°53'03"	E025°29'45"	57,56	14,86
GR13DB04	Sporades, Skyros, Magazia	N38°54'46"	E024°34'18"		
TU08DB01	Bodrum, Gümbet	N37°01'52.54"	E027°24'17.62"	79,94	7,74
CY09DB01	Xeros, Lefka, jun 2005	N35°08'34"	E032°50'19"	51,70	7,76
TN09DB01	Tunisia			88,51	0,24
TN10DB02	Douz, 2002				
LI09DB01	Great Sand Sea			96,39	1,17
LI09DB02	Erg Murzuq	N25°54'	E13°54'	98,06	1,18
LI09DB03	Erg Awbari	N26°35'	E12°45'	98,27	1,14
LI09DB04	Takharkhuri, Fezzan, feb 2003			97,54	1,43
LI09DB05	Al Aweinat (Serdeles), Fezzan, feb 2003			89,58	2,79
EG08DB01	Great Sand Sea			97,53	1,45
EG08DB02	Selina Sand Sheets			98,17	1,04
EG08DB03	Selina Sand Sheets			99,00	0,82
EG08DB04	Wadi Abdel Malek			98,49	0,59
EG08DB05	Alexandria			39,13	0,30
EG09DB06	Makadi Bay	N26°59'	E033°54'	64,07	9,65
EG09DB09	Mount Sinai, jul 1987	N28°32'	E033°58'	72,24	13,78
EG09DB10	Saqqara, near pyramids, mar 2000	N29°52'	E031°13'	84,91	2,62
EG09DB11	Saqqara, near pyramids, apr 2006	N29°52'	E031°13'	90,62	1,62
EG09DB12	Giza, Cairo, near pyramids, apr 2006	N29°58'	E031°07'	58,02	2,59
EG09DB13	Aswan, Nile banks, apr 2006	N24°04'	E032°51'	95,83	1,28
EG10DB14	Great Sand Sea between Ain Della and Farafra	N27°13'	E027°28'	93,83	0,50
EG10DB15	Abu Ballas	N24°26'	E027°38'	95,45	1,04
EG10DB16	Water Mountain	N25°23'	E028°17'	94,52	0,32
EG10DB17	El Gouna	N27°23'	E033°40'		
EG13DB18	Amarna	N27°38'50"	E030°54'53"		
IS08DB01	River Belus, Israel, Levant (Brill 674; Degryse and Schneider, 2008)	N32°54'32"	E35°04'53"		
IS08DB02	River Belus, Israel, Levant (Brill 679; Degryse and Schneider, 2008)	N32°54'32"	E35°04'53"		
IS08DB03	River Belus, Israel, Levant (Brill 681; Degryse and Schneider, 2008)	N32°54'17"	E35°04'52"		
IS09DB04	Akko, beach, aug 1979	N32°54'46"	E35°04'52"	2,61	0,43
IS09DB05	Haifa, beach, aug 1979	N32°47'09"	E34°57'17"	3,34	0,33
IS09DB06	Masada, below the fortress, jul 1987	N31°19'05"	E35°21'18"	10,76	1,80
IS09DB07	Red Sea, near Eilat, jul 1987	N29°32'53"	E34°57'49"	63,81	12,72
IS09DB08	Dead Sea, near Ein Gedi, jul 1987	N31°27'27"	E35°23'57"	81,12	1,16

Fe$_2$O$_3$(t)	MgO	MnO	CaO	Na$_2$O	K$_2$O	TiO$_2$	P$_2$O$_5$	LOI	Total Oxides
(Wt.%)	(Wt.%)	(Wt.%)	(Wt.%)	(Wt.%)	(Wt.%)	(Wt.%)	(Wt.%)	(Wt.%)	
ICP-OES	ICP-OES	ICP-OES	ICP-OES	ICP-OES	ICP-OES	ICP-OES	ICP-OES		
0,22	0,07	0,01	2,64	0,20	0,67	0,04	0,02	2,55	96,80
1,76	1,45	0,11	21,78	0,40	0,34	0,83	0,04	16,34	98,97
0,55	0,78	0,06	43,68	0,17	0,09	0,05	0,02	35,33	98,66
3,11	1,52	0,05	9,05	2,66	5,25	0,40	0,17	5,48	100,10
1,19	0,67	0,02	2,94	1,87	2,83	0,15	0,05	3,30	100,72
6,00	11,07	0,13	10,00	1,26	0,87	0,47	0,06	11,15	100,47
0,26	0,12	0,01	6,66	0,04	0,06	0,04	0,01	5,29	101,24
0,29	0,07	0,01	0,35	0,16	0,41	0,12	0,01	0,60	99,59
0,30	0,01	0,00	0,05	0,04	0,08	0,09	0,01	0,61	100,42
0,34	0,04	0,00	0,03	0,05	0,17	0,07	0,01	0,76	100,89
0,46	0,04	0,01	0,10	0,05	0,19	0,08	0,01	0,83	100,75
1,17	0,41	0,03	2,01	0,54	0,40	0,15	0,03	3,41	100,52
0,19	0,03	0,01	0,06	0,16	0,54	0,09	0,01	0,40	100,46
0,21	0,02	0,00	0,05	0,10	0,29	0,06	0,01	0,41	100,36
0,20	0,00	0,00	0,06	0,05	0,17	0,04	0,12	0,36	100,83
0,51	0,01	0,00	0,03	0,03	0,04	0,08	0,01	0,52	100,32
0,33	2,25	0,03	30,99	0,35	0,09	0,03	0,09	27,48	101,08
8,24	0,95	0,12	5,85	3,33	2,81	1,48	0,10	5,29	101,89
1,70	0,40	0,05	1,42	3,87	4,57	0,24	0,04	2,43	100,75
0,84	0,28	0,02	5,23	0,49	0,62	0,23	0,06	5,04	100,34
0,55	0,14	0,01	2,63	0,23	0,40	0,15	0,06	2,51	98,92
1,39	0,59	0,04	18,84	0,60	0,63	0,28	0,11	16,04	99,12
0,29	0,06	0,01	1,24	0,10	0,29	0,10	0,01	1,19	100,40
0,34	0,09	0,00	0,92	0,05	0,11	0,04	0,02	1,04	96,94
0,31	0,05	0,01	0,25	0,04	0,11	0,12	0,03	0,68	98,08
0,56	0,07	0,01	1,31	0,04	0,04	0,03	0,74	0,64	98,27
0,70	3,39	0,05	48,77	0,39	0,06	0,04	0,07	42,54	99,05
0,36	3,17	0,04	49,31	0,57	0,06	0,03	0,04	43,06	100,32
0,97	5,90	0,02	36,41	1,69	0,50	0,17	0,39	38,17	96,78
2,67	1,25	0,04	7,06	3,64	3,72	0,50	0,10	5,46	100,98
0,17	2,36	0,01	6,17	0,09	0,36	0,08	0,13	7,27	98,91

Sample	Beach location	Latitude (°N)	Longitude (°E)	SiO$_2$	Al$_2$O$_3$
				(Wt.%)	(Wt.%)
				ICP-OES	ICP-OES
IS13DB09	Tel Dor	N32°36'52"	E034°55'03"		
IS13DB10	Tel Dor	N32°36'58"	E034°54'58"		
IS13DB12	Shikmona beach, west of Tel Shikmona	N32°49'32"	E034°57'19"		
IS13DB13	Nitsanim beach, between Ashdod and Ashkelon	N31°44'37"	E034°35'57"		
OM10DB01	Saih Rawl (oilwell #29)	N21°20'	E056°40'	64,78	4,07
BE08DB01	Bastion, Antwerpen, Belgium				
NE08DB01	Hulst			92,23	3,52
UK08DB01	Grandcourt Farm, Kings Lynn Quarry, Norfolk			93,72	1,21
UK08DB02	Wicken South, Kings Lynn Quarry, Norfolk			94,60	1,58

Fe$_2$O$_3$(t)	MgO	MnO	CaO	Na$_2$O	K$_2$O	TiO$_2$	P$_2$O$_5$	LOI	Total Oxides
(Wt.%)	(Wt.%)	(Wt.%)	(Wt.%)	(Wt.%)	(Wt.%)	(Wt.%)	(Wt.%)	(Wt.%)	
ICP-OES	ICP-OES	ICP-OES	ICP-OES	ICP-OES	ICP-OES	ICP-OES	ICP-OES		
1,65	2,25	0,04	12,10	0,88	1,17	0,21	0,03	10,81	98,01
0,50	0,14	0,01	0,31	0,72	1,24	0,12	0,05	0,86	99,70
0,56	0,03	0,00	0,03	0,05	0,69	0,03	0,02	0,73	97,07
0,63	0,04	0,00	0,06	0,05	0,65	0,03	0,03	0,65	98,33

Sample	As	Ba	Co	Cr	Cu	La	Nd	Ni
	(ppm)	(ppm)	(ppm)	(ppm)	(ppm)	(ppm)	(ppm)	(ppm)
	ICP-OES	ICP-OES	ICP-OES	ICP-OES	ICP-OES	ICP-OES	ICP-OES	ICP-OES
Spain								
SP10DB46	bdl.	134,5	5,0	428,3	24,9	bdl.	19,9	74,7
SP10DB45	44,9	79,8	5,0	423,7	19,9	bdl.	10,0	94,7
SP10DB44	bdl.	49,8	5,0	298,5	10,0	bdl.	14,9	59,7
SP10DB43	35,0	105,1	5,0	160,2	10,0	bdl.	10,0	50,1
SP10DB42	bdl.	74,6	10,0	228,9	19,9	bdl.	bdl.	59,7
SP10DB41	34,8	49,7	5,0	89,4	29,8	bdl.	bdl.	19,9
SP10DB40	bdl.	10,0	5,0	114,7	29,9	bdl.	bdl.	44,9
SP10DB39	bdl.	14,9	5,0	114,5	14,9	bdl.	bdl.	39,8
SP10DB38	25,0	20,0	5,0	234,5	5,0	bdl.	bdl.	69,9
SP10DB37	bdl.	39,8	5,0	488,0	24,9	bdl.	5,0	99,6
SP10DB36	bdl.	119,9	10,0	944,1	10,0	15,0	5,0	154,8
SP10DB35	bdl.	155,0	15,0	1045,0	25,0	10,0	25,0	330,0
SP10DB34	bdl.	119,5	14,9	971,1	34,9	10,0	19,9	403,4
SP10DB33	bdl.	209,2	14,9	597,6	14,9	14,9	14,9	199,2
SP10DB32	bdl.	175,2	10,0	515,5	10,0	15,0	bdl.	165,2
SP10DB31	39,9	194,6	10,0	214,6	34,9	29,9	10,0	69,9
SP10DB30	bdl.	89,8	10,0	234,5	10,0	25,0	bdl.	39,9
SP10DB29	bdl.	153,9	5,0	223,4	9,9	9,9	24,8	79,4
SP10DB28	bdl.	109,5	5,0	358,2	24,9	10,0	bdl.	99,5
SP10DB27	bdl.	84,8	5,0	304,4	20,0	5,0	bdl.	104,8
SP10DB26	bdl.	100,0	5,0	255,0	5,0	5,0	bdl.	70,0
SP10DB25	bdl.	49,7	5,0	159,0	5,0	5,0	bdl.	44,7
SP10DB24	bdl.	125,1	20,0	225,2	5,0	20,0	bdl.	45,0
SP10DB23	bdl.	1564,1	9,9	168,8	24,8	14,9	9,9	49,7
SP10DB22	bdl.	114,9	5,0	354,6	10,0	5,0	30,0	104,9
SP10DB21	bdl.	407,6	9,9	213,7	24,9	14,9	9,9	64,6
SP10DB20	bdl.	79,7	5,0	268,9	5,0	bdl.	14,9	54,8
SP10DB19	bdl.	204,8	10,0	179,8	35,0	10,0	bdl.	54,9
SP10DB18	bdl.	45,0	bdl.	109,9	10,0	5,0	bdl.	30,0
SP08DB01	5,0	75,0	5,0	60,0	5,0	5,0	45,0	15,0
SP10DB17	bdl.	44,9	5,0	29,9	5,0	5,0	bdl.	bdl.
SP10DB16	bdl.	94,6	5,0	89,6	5,0	bdl.	bdl.	14,9
SP10DB15	bdl.	15,0	5,0	10,0	bdl.	5,0	bdl.	bdl.
SP10DB14	bdl.	25,0	5,0	39,9	5,0	5,0	bdl.	bdl.
SP10DB13	bdl.	99,9	5,0	339,7	10,0	5,0	bdl.	84,9
SP10DB12	bdl.	24,9	5,0	84,7	5,0	bdl.	bdl.	19,9
SP10DB11	bdl.	40,0	5,0	290,3	20,0	bdl.	bdl.	65,1
SP10DB10	bdl.	74,8	5,0	204,4	19,9	5,0	bdl.	34,9
SP10DB09	bdl.	39,9	5,0	164,7	5,0	bdl.	bdl.	39,9
SP10DB08	bdl.	54,6	5,0	292,9	19,9	bdl.	bdl.	74,5
SP10DB07	bdl.	54,9	5,0	149,7	10,0	5,0	bdl.	25,0

Pb (ppm) ICP-OES	Rb (ppm) ICP-OES	Sb (ppm) ICP-OES	Sc (ppm) ICP-OES	Sn (ppm) ICP-OES	Sr (ppm) ICP-OES	V (ppm) ICP-OES	W (ppm) ICP-OES	Zn (ppm) ICP-OES	Zr (ppm) ICP-OES
bdl.	nd.	bdl.	nd.	bdl.	79,7	19,9	bdl.	34,9	34,9
54,8	nd.	bdl.	nd.	10,0	89,7	10,0	bdl.	44,9	15,0
bdl.	nd.	bdl.	nd.	bdl.	34,8	10,0	bdl.	bdl.	19,9
30,0	nd.	bdl.	nd.	10,0	75,1	5,0	25,0	bdl.	10,0
49,8	nd.	bdl.	nd.	5,0	293,5	bdl.	bdl.	10,0	19,9
69,5	nd.	bdl.	nd.	bdl.	764,6	bdl.	14,9	5,0	34,8
79,8	nd.	bdl.	nd.	bdl.	668,0	bdl.	10,0	bdl.	34,9
bdl.	nd.	bdl.	nd.	5,0	667,3	bdl.	10,0	bdl.	10,0
bdl.	nd.	bdl.	nd.	5,0	384,2	5,0	bdl.	bdl.	20,0
29,9	nd.	bdl.	nd.	bdl.	84,7	5,0	bdl.	10,0	64,7
25,0	nd.	bdl.	nd.	bdl.	69,9	45,0	bdl.	74,9	69,9
55,0	nd.	bdl.	nd.	bdl.	195,0	50,0	20,0	40,0	55,0
59,8	nd.	bdl.	nd.	bdl.	214,1	34,9	bdl.	39,8	49,8
bdl.	nd.	bdl.	nd.	bdl.	109,6	49,8	29,9	39,8	84,7
bdl.	nd.	bdl.	nd.	5,0	130,1	50,1	10,0	30,0	80,1
124,8	nd.	bdl.	nd.	bdl.	114,8	69,9	29,9	119,8	159,7
bdl.	nd.	bdl.	nd.	bdl.	119,8	54,9	10,0	374,3	129,7
79,4	nd.	bdl.	nd.	bdl.	104,3	34,8	19,9	34,8	109,2
54,7	nd.	bdl.	nd.	bdl.	74,6	34,8	24,9	94,5	149,3
54,9	nd.	bdl.	nd.	5,0	209,6	20,0	54,9	20,0	64,9
55,0	nd.	bdl.	nd.	bdl.	90,0	25,0	5,0	20,0	65,0
54,7	nd.	bdl.	nd.	bdl.	462,2	24,9	14,9	19,9	34,8
65,1	nd.	bdl.	nd.	bdl.	175,2	215,2	bdl.	60,1	45,0
292,9	nd.	bdl.	nd.	bdl.	253,2	29,8	19,9	109,2	89,4
104,9	nd.	bdl.	nd.	bdl.	99,9	15,0	15,0	25,0	69,9
357,9	nd.	bdl.	nd.	bdl.	139,2	34,8	24,9	99,4	89,5
24,9	nd.	bdl.	nd.	bdl.	29,9	14,9	10,0	14,9	54,8
1398,6	nd.	bdl.	nd.	bdl.	204,8	40,0	20,0	789,2	69,9
64,9	nd.	bdl.	nd.	bdl.	414,6	5,0	bdl.	20,0	20,0
bdl.	20,0	bdl.	5,0	bdl.	390,0	20,0	bdl.	bdl.	45,0
19,9	nd.	bdl.	nd.	bdl.	204,4	bdl.	5,0	bdl.	5,0
bdl.	nd.	bdl.	nd.	bdl.	278,9	bdl.	bdl.	bdl.	39,8
bdl.	nd.	bdl.	nd.	5,0	1105,0	bdl.	bdl.	bdl.	5,0
39,9	nd.	bdl.	nd.	bdl.	289,4	bdl.	5,0	bdl.	25,0
59,9	nd.	bdl.	nd.	bdl.	354,6	bdl.	bdl.	bdl.	20,0
39,9	nd.	bdl.	nd.	5,0	1311,1	bdl.	bdl.	bdl.	10,0
bdl.	nd.	bdl.	nd.	bdl.	205,2	bdl.	bdl.	bdl.	15,0
bdl.	nd.	bdl.	nd.	bdl.	184,4	bdl.	bdl.	bdl.	19,9
bdl.	nd.	bdl.	nd.	bdl.	134,7	bdl.	bdl.	bdl.	39,9
59,6	nd.	bdl.	nd.	bdl.	173,8	bdl.	bdl.	9,9	14,9
49,9	nd.	bdl.	nd.	bdl.	174,7	bdl.	bdl.	5,0	29,9

Sample	As	Ba	Co	Cr	Cu	La	Nd	Ni
	(ppm)	(ppm)	(ppm)	(ppm)	(ppm)	(ppm)	(ppm)	(ppm)
	ICP-OES	ICP-OES	ICP-OES	ICP-OES	ICP-OES	ICP-OES	ICP-OES	ICP-OES
SP10DB06	bdl.	94,8	5,0	139,7	5,0	10,0	bdl.	20,0
SP08DB02	bdl.	100,0	10,0	35,0	bdl.	30,0	35,0	15,0
SP10DB05	bdl.	74,7	5,0	164,3	5,0	39,8	bdl.	29,9
SP10DB04	bdl.	54,9	5,0	279,7	5,0	bdl.	bdl.	64,9
SP08FH05	15,0	79,8	5,0	179,6	5,0	29,9	15,0	44,9
SP08FH06	bdl.	134,2	5,0	188,9	5,0	5,0	19,9	44,7
SP08FH07	bdl.	104,2	5,0	173,6	5,0	24,8	14,9	39,7
SP08FH08	bdl.	438,7	5,0	294,1	5,0	10,0	10,0	64,8
SP10DB47	bdl.	263,9	5,0	114,5	54,8	5,0	10,0	14,9
SP08FH04	bdl.	427,0	5,0	302,9	9,9	5,0	9,9	79,4
SP08FH03	bdl.	504,0	5,0	359,3	10,0	39,9	49,9	84,8
SP08FH01	bdl.	422,9	5,0	447,8	10,0	5,0	24,9	104,5
SP08FH02	bdl.	471,2	5,0	312,5	9,9	5,0	19,8	69,4
SP08FH12	9,9	556,7	9,9	477,1	9,9	bdl.	19,9	119,3
SP08FH11	bdl.	521,9	5,0	447,3	9,9	bdl.	19,9	104,4
SP08FH10	bdl.	586,5	5,0	328,0	5,0	bdl.	19,9	74,6
SP08FH09	bdl.	462,7	5,0	318,4	5,0	124,4	84,6	69,7
SP09DB03	30,0	595,0	5,0	75,0	30,0	25,0	20,0	35,0
France								
FR09DB02	bdl.	50,0	10,0	10,0	bdl.	bdl.	30,0	10,0
FR09DB03	5,0	485,0	5,0	25,0	10,0	5,0	20,0	15,0
FR09DB04	bdl.	460,0	10,0	20,0	5,0	bdl.	20,0	10,0
FR09DB05	15,0	225,0	10,0	25,0	5,0	15,0	30,0	10,0
FR09DB06	bdl.	375,0	5,0	145,0	5,0	5,0	25,0	50,0
FR09DB07	20,0	375,0	10,0	50,0	10,0	10,0	25,0	15,0
FR09DB08	45,0	225,0	5,0	35,0	5,0	5,0	30,0	15,0
FR09DB09	10,0	250,0	10,0	20,0	bdl.	bdl.	40,0	10,0
FR09DB10	20,0	265,0	10,0	25,0	5,0	5,0	25,0	10,0
FR09DB11	20,0	175,0	20,0	150,0	bdl.	110,0	65,0	60,0
FR09DB12	bdl.	35,0	10,0	10,0	bdl.	bdl.	40,0	10,0
FR09DB13	bdl.	350,0	5,0	10,0	bdl.	bdl.	20,0	10,0
FR09DB14	bdl.	75,0	5,0	5,0	bdl.	bdl.	45,0	bdl.
FR09DB16	bdl.	175,0	10,0	15,0	10,0	bdl.	20,0	10,0
FR09DB17	15,0	175,0	5,0	25,0	bdl.	bdl.	20,0	5,0
FR09DB18	bdl.	355,0	10,0	10,0	5,0	bdl.	25,0	bdl.
FR09DB19	40,0	385,0	10,0	10,0	5,0	bdl.	25,0	5,0
FR09DB20	bdl.	235,0	10,0	20,0	5,0	15,0	40,0	5,0
FR09DB01	bdl.	285,0	10,0	20,0	5,0	bdl.	35,0	15,0
Italy								
IT09DB12	bdl.	100,0	10,0	15,0	5,0	bdl.	30,0	5,0
IT09DB13	20,0	150,0	10,0	10,0	5,0	bdl.	25,0	bdl.
IT09DB14	bdl.	300,0	10,0	175,0	10,0	10,0	20,0	35,0
IT09DB15	bdl.	190,0	10,0	545,0	10,0	5,0	20,0	130,0
IT09DB16a	bdl.	135,0	50,0	1835,0	45,0	20,0	25,0	565,0

Pb (ppm) ICP-OES	Rb (ppm) ICP-OES	Sb (ppm) ICP-OES	Sc (ppm) ICP-OES	Sn (ppm) ICP-OES	Sr (ppm) ICP-OES	V (ppm) ICP-OES	W (ppm) ICP-OES	Zn (ppm) ICP-OES	Zr (ppm) ICP-OES
bdl.	nd.	bdl.	nd.	bdl.	229,5	5,0	bdl.	5,0	49,9
bdl.	20,0	bdl.	5,0	bdl.	225,0	25,0	bdl.	bdl.	165,0
49,8	nd.	bdl.	nd.	bdl.	214,1	24,9	bdl.	19,9	174,3
40,0	nd.	10,0	nd.	bdl.	279,7	bdl.	bdl.	bdl.	20,0
15,0	5,0	10,0	5,0	bdl.	354,3	34,9	bdl.	25,0	49,9
bdl.	39,8	bdl.	5,0	bdl.	233,6	14,9	bdl.	bdl.	49,7
bdl.	19,8	5,0	5,0	bdl.	292,7	24,8	bdl.	bdl.	79,4
10,0	74,8	5,0	5,0	5,0	149,6	19,9	bdl.	bdl.	54,8
84,7	nd.	bdl.	nd.	bdl.	144,4	5,0	bdl.	10,0	14,9
14,9	64,5	bdl.	5,0	bdl.	153,9	9,9	bdl.	19,9	59,6
bdl.	89,8	bdl.	5,0	bdl.	159,7	5,0	bdl.	bdl.	39,9
10,0	69,7	10,0	5,0	bdl.	109,5	bdl.	bdl.	bdl.	44,8
bdl.	89,3	5,0	5,0	bdl.	94,2	bdl.	bdl.	9,9	44,6
19,9	94,4	5,0	bdl.	bdl.	69,6	9,9	bdl.	bdl.	54,7
9,9	79,5	9,9	bdl.	bdl.	69,6	bdl.	bdl.	bdl.	44,7
bdl.	114,3	9,9	bdl.	bdl.	99,4	bdl.	bdl.	bdl.	44,7
bdl.	74,6	5,0	5,0	bdl.	104,5	14,9	bdl.	bdl.	49,8
15,0	100,0	10,0	10,0	bdl.	125,0	95,0	bdl.	100,0	150,0
bdl.	bdl.	bdl.	bdl.	bdl.	360,0	bdl.	bdl.	bdl.	25,0
bdl.	130,0	bdl.	5,0	bdl.	80,0	25,0	bdl.	bdl.	80,0
20,0	115,0	5,0	5,0	bdl.	85,0	20,0	bdl.	bdl.	65,0
bdl.	55,0	25,0	5,0	bdl.	365,0	20,0	bdl.	bdl.	55,0
135,0	100,0	bdl.	5,0	bdl.	180,0	25,0	bdl.	bdl.	60,0
30,0	110,0	bdl.	5,0	bdl.	200,0	30,0	bdl.	bdl.	55,0
bdl.	50,0	5,0	5,0	bdl.	335,0	15,0	bdl.	bdl.	85,0
bdl.	55,0	15,0	5,0	bdl.	575,0	10,0	bdl.	bdl.	35,0
bdl.	55,0	10,0	5,0	bdl.	215,0	15,0	bdl.	bdl.	80,0
25,0	50,0	15,0	15,0	bdl.	340,0	70,0	bdl.	35,0	320,0
bdl.	15,0	bdl.	5,0	bdl.	880,0	bdl.	bdl.	bdl.	15,0
bdl.	35,0	10,0	bdl.	bdl.	70,0	bdl.	bdl.	bdl.	35,0
15,0	10,0	bdl.	bdl.	bdl.	215,0	bdl.	bdl.	bdl.	10,0
105,0	40,0	10,0	5,0	5,0	50,0	15,0	bdl.	155,0	80,0
bdl.	35,0	5,0	5,0	bdl.	135,0	25,0	bdl.	5,0	100,0
bdl.	85,0	bdl.	5,0	10,0	145,0	5,0	bdl.	bdl.	50,0
40,0	65,0	5,0	5,0	bdl.	105,0	5,0	bdl.	bdl.	65,0
20,0	50,0	20,0	5,0	bdl.	400,0	15,0	bdl.	bdl.	50,0
bdl.	80,0	15,0	5,0	bdl.	355,0	bdl.	bdl.	bdl.	55,0
bdl.	20,0	5,0	5,0	bdl.	550,0	5,0	bdl.	bdl.	55,0
55,0	30,0	25,0	5,0	10,0	75,0	10,0	bdl.	bdl.	40,0
40,0	50,0	bdl.	10,0	bdl.	100,0	55,0	bdl.	bdl.	175,0
10,0	25,0	10,0	10,0	bdl.	100,0	50,0	bdl.	bdl.	60,0
95,0	20,0	bdl.	15,0	15,0	125,0	125,0	10,0	145,0	90,0

Sample	As	Ba	Co	Cr	Cu	La	Nd	Ni
	(ppm)	(ppm)	(ppm)	(ppm)	(ppm)	(ppm)	(ppm)	(ppm)
	ICP-OES	ICP-OES	ICP-OES	ICP-OES	ICP-OES	ICP-OES	ICP-OES	ICP-OES
IT09DB16b	bdl.	110,0	40,0	1790,0	25,0	15,0	25,0	730,0
IT09DB17	bdl.	125,0	30,0	720,0	250,0	10,0	25,0	510,0
IT09DB18	bdl.	75,0	35,0	1790,0	20,0	10,0	20,0	605,0
IT09DB19	bdl.	315,0	5,0	190,0	10,0	10,0	25,0	125,0
IT09DB09	bdl.	265,0	10,0	290,0	5,0	15,0	30,0	30,0
IT10DB30	15,0	294,7	5,0	269,7	bdl.	bdl.	20,0	79,9
IT10DB31	bdl.	274,2	5,0	353,9	5,0	5,0	15,0	89,7
IT10DB32	25,0	25,0	10,0	738,5	bdl.	bdl.	15,0	239,5
IT10DB33	bdl.	69,9	25,0	1108,9	15,0	bdl.	15,0	454,5
IT08DB05	bdl.	260,0	10,0	125,0	10,0	10,0	20,0	40,0
IT10DB34	bdl.	129,5	5,0	293,8	bdl.	bdl.	14,9	59,8
IT08DB01	20,0	140,0	10,0	315,0	bdl.	5,0	25,0	10,0
IT10DB35	bdl.	134,7	10,0	479,0	bdl.	5,0	15,0	64,9
IT10DB36	14,9	133,9	9,9	257,9	bdl.	5,0	14,9	69,4
IT10DB37	10,0	154,4	5,0	149,4	5,0	5,0	14,9	69,7
IT10DB38	34,9	748,5	5,0	144,7	bdl.	44,9	29,9	49,9
IT10DB39	bdl.	1253,7	10,0	274,7	bdl.	50,0	35,0	74,9
IT10DB40	bdl.	1003,0	19,9	263,2	bdl.	34,8	34,8	59,6
IT10DB41	9,9	312,8	34,8	337,6	bdl.	69,5	69,5	79,4
IT10DB42	bdl.	189,1	34,8	527,4	bdl.	39,8	54,7	129,4
IT08DB06	bdl.	190,0	30,0	405,0	bdl.	70,0	70,0	70,0
IT10DB43	bdl.	294,1	5,0	224,3	bdl.	15,0	15,0	49,9
IT10DB44	15,0	209,4	5,0	189,4	bdl.	bdl.	15,0	39,9
IT09DB22	nd.	nd.	nd.	nd.	nd.	nd.	nd.	nd.
IT10DB45	bdl.	379,2	10,0	159,7	bdl.	44,9	25,0	25,0
IT09DB23	10,0	320,0	5,0	100,0	bdl.	30,0	40,0	20,0
IT09DB21	nd.	nd.	nd.	nd.	nd.	nd.	nd.	nd.
IT09DB25	20,0	500,0	5,0	75,0	5,0	15,0	30,0	15,0
IT09DB26	20,0	435,0	5,0	55,0	5,0	20,0	35,0	5,0
IT09DB27	10,0	535,0	5,0	30,0	5,0	10,0	30,0	15,0
IT09DB20	nd.	nd.	nd.	nd.	nd.	nd.	nd.	nd.
IT09DB28	30,0	420,0	10,0	20,0	5,0	10,0	35,0	10,0
IT09DB24	25,0	575,0	5,0	25,0	5,0	25,0	30,0	10,0
IT08DB02	bdl.	650,0	15,0	145,0	20,0	20,0	45,0	40,0
IT10DB46	bdl.	119,6	10,0	124,6	bdl.	bdl.	19,9	29,9
IT10DB47	bdl.	278,6	5,0	318,4	bdl.	bdl.	19,9	79,6
IT10DB48	bdl.	69,8	5,0	119,6	10,0	5,0	15,0	54,8
IT10DB49	bdl.	19,9	5,0	109,5	bdl.	bdl.	14,9	39,8
IT10DB50	bdl.	333,3	19,9	288,6	10,0	14,9	19,9	84,6
IT10DB51	bdl.	472,2	9,9	223,7	9,9	19,9	34,8	79,5
IT10DB52	bdl.	1255,0	5,0	373,5	bdl.	14,9	24,9	104,6
IT10DB53	9,9	764,6	5,0	153,9	bdl.	19,9	34,8	24,8
IT10DB54	bdl.	495,0	5,0	320,0	5,0	bdl.	20,0	80,0
IT10DB55	5,0	587,6	5,0	139,4	bdl.	5,0	19,9	24,9

Pb (ppm) ICP-OES	Rb (ppm) ICP-OES	Sb (ppm) ICP-OES	Sc (ppm) ICP-OES	Sn (ppm) ICP-OES	Sr (ppm) ICP-OES	V (ppm) ICP-OES	W (ppm) ICP-OES	Zn (ppm) ICP-OES	Zr (ppm) ICP-OES
35,0	25,0	bdl.	15,0	bdl.	95,0	80,0	bdl.	120,0	85,0
45,0	25,0	bdl.	15,0	bdl.	70,0	75,0	bdl.	145,0	60,0
25,0	20,0	bdl.	15,0	bdl.	145,0	75,0	10,0	55,0	40,0
15,0	65,0	bdl.	10,0	bdl.	275,0	35,0	bdl.	15,0	60,0
25,0	50,0	bdl.	5,0	bdl.	250,0	25,0	bdl.	bdl.	100,0
10,0	129,9	bdl.	5,0	bdl.	204,8	20,0	bdl.	25,0	54,9
15,0	bdl.	bdl.	5,0	bdl.	259,2	19,9	bdl.	19,9	49,9
bdl.	bdl.	15,0	10,0	bdl.	718,6	34,9	bdl.	44,9	20,0
20,0	84,9	5,0	10,0	bdl.	264,7	50,0	bdl.	59,9	40,0
135,0	130,0	10,0	10,0	5,0	155,0	60,0	bdl.	10,0	75,0
14,9	bdl.	10,0	5,0	bdl.	224,1	5,0	bdl.	24,9	29,9
5,0	25,0	15,0	5,0	bdl.	60,0	20,0	bdl.	0,0	55,0
20,0	bdl.	10,0	10,0	bdl.	374,3	29,9	bdl.	39,9	69,9
19,8	19,8	9,9	9,9	bdl.	391,9	24,8	bdl.	44,6	49,6
19,9	bdl.	5,0	10,0	bdl.	453,2	29,9	bdl.	49,8	39,8
39,9	194,6	10,0	10,0	5,0	863,3	39,9	bdl.	69,9	84,8
59,9	254,7	15,0	20,0	5,0	1188,8	94,9	bdl.	45,0	159,8
34,8	193,6	14,9	34,8	5,0	993,0	109,2	bdl.	34,8	153,9
14,9	99,3	9,9	49,7	bdl.	397,2	258,2	bdl.	54,6	278,1
19,9	bdl.	5,0	54,7	bdl.	417,9	213,9	bdl.	44,8	273,6
bdl.	25,0	bdl.	45,0	bdl.	405,0	250,0	bdl.	45,0	335,0
24,9	bdl.	15,0	10,0	bdl.	373,9	39,9	bdl.	19,9	104,7
15,0	114,7	5,0	5,0	bdl.	478,6	10,0	bdl.	bdl.	29,9
nd.	nd.	nd.	nd.	nd.	nd.	nd.	nd.	nd.	nd.
25,0	bdl.	bdl.	20,0	bdl.	429,1	94,8	bdl.	25,0	124,8
5,0	40,0	15,0	15,0	bdl.	380,0	75,0	bdl.	bdl.	125,0
nd.	nd.	nd.	nd.	nd.	nd.	nd.	nd.	nd.	nd.
bdl.	75,0	25,0	15,0	5,0	340,0	55,0	bdl.	bdl.	75,0
bdl.	75,0	15,0	10,0	5,0	335,0	65,0	bdl.	10,0	205,0
15,0	105,0	bdl.	5,0	bdl.	390,0	30,0	bdl.	bdl.	75,0
nd.	nd.	nd.	nd.	nd.	nd.	nd.	nd.	nd.	nd.
20,0	65,0	35,0	5,0	bdl.	385,0	20,0	bdl.	bdl.	50,0
30,0	130,0	15,0	5,0	bdl.	470,0	30,0	bdl.	bdl.	90,0
40,0	85,0	bdl.	25,0	bdl.	515,0	125,0	bdl.	bdl.	110,0
24,9	bdl.	15,0	15,0	bdl.	229,3	44,9	bdl.	24,9	59,8
19,9	124,4	10,0	5,0	bdl.	139,3	19,9	bdl.	24,9	49,8
15,0	bdl.	5,0	5,0	5,0	229,3	24,9	bdl.	54,8	89,7
bdl.	bdl.	bdl.	5,0	bdl.	154,2	14,9	bdl.	10,0	24,9
24,9	bdl.	5,0	19,9	bdl.	223,9	114,4	bdl.	64,7	139,3
9,9	149,1	9,9	9,9	bdl.	119,3	74,6	bdl.	59,6	109,3
29,9	119,5	10,0	5,0	bdl.	234,1	10,0	bdl.	bdl.	94,6
14,9	134,1	9,9	5,0	bdl.	178,7	14,9	bdl.	9,9	94,3
20,0	140,0	10,0	5,0	bdl.	170,0	20,0	bdl.	5,0	60,0
29,9	149,4	10,0	5,0	bdl.	154,4	19,9	bdl.	5,0	49,8

Sample	As	Ba	Co	Cr	Cu	La	Nd	Ni
	(ppm)	(ppm)	(ppm)	(ppm)	(ppm)	(ppm)	(ppm)	(ppm)
	ICP-OES	ICP-OES	ICP-OES	ICP-OES	ICP-OES	ICP-OES	ICP-OES	ICP-OES
IT10DB56	10,0	384,2	5,0	394,2	5,0	5,0	15,0	109,8
IT10DB57	bdl.	164,3	5,0	259,0	5,0	5,0	10,0	89,6
IT10DB58	14,9	333,0	9,9	233,6	9,9	14,9	9,9	74,6
IT10DB59	5,0	34,9	5,0	283,9	bdl.	bdl.	10,0	74,7
IT10DB60	15,0	45,0	5,0	290,0	5,0	bdl.	20,0	30,0
IT10DB61	bdl.	10,0	5,0	89,6	bdl.	bdl.	bdl.	34,9
IT10DB62	bdl.	69,9	5,0	314,4	bdl.	bdl.	15,0	64,9
IT10DB63	10,0	59,8	5,0	383,8	5,0	bdl.	10,0	74,8
IT10DB64	14,9	59,6	5,0	139,0	bdl.	bdl.	9,9	44,7
IT10DB65	bdl.	19,9	5,0	194,4	5,0	bdl.	10,0	54,8
IT10DB66	15,0	164,5	5,0	149,6	bdl.	bdl.	10,0	34,9
IT08DB08	bdl.	270,0	10,0	10,0	5,0	bdl.	35,0	0,0
IT10DB67	bdl.	175,0	5,0	180,0	bdl.	bdl.	10,0	50,0
IT10DB68	10,0	139,7	5,0	119,8	bdl.	bdl.	10,0	29,9
IT10DB69	bdl.	124,1	5,0	307,8	bdl.	bdl.	9,9	79,4
IT10DB70	9,9	19,9	5,0	24,8	bdl.	bdl.	19,9	5,0
IT10DB71	bdl.	869,8	9,9	313,1	5,0	5,0	19,9	79,5
IT10DB72	19,9	333,7	10,0	293,8	14,9	19,9	19,9	94,6
IT10DB73	19,9	392,6	9,9	164,0	9,9	24,9	29,8	59,6
IT10DB74	bdl.	446,9	5,0	268,1	9,9	14,9	19,9	84,4
IT10DB75	bdl.	435,0	5,0	270,0	15,0	20,0	25,0	80,0
IT10DB76	bdl.	434,1	10,0	259,5	5,0	20,0	25,0	84,8
IT10DB77	bdl.	610,0	5,0	245,0	10,0	5,0	15,0	70,0
IT10DB78	9,9	591,5	5,0	218,7	bdl.	9,9	24,9	59,6
IT10DB79	10,0	640,0	5,0	355,0	10,0	20,0	25,0	85,0
IT10DB80	bdl.	805,0	5,0	290,0	5,0	40,0	40,0	70,0
IT10DB81	bdl.	657,4	5,0	333,7	10,0	19,9	24,9	74,7
IT10DB82	bdl.	419,6	5,0	239,8	10,0	10,0	20,0	74,9
IT10DB83	bdl.	473,6	10,0	254,2	10,0	19,9	19,9	59,8
IT10DB84	bdl.	84,6	5,0	457,7	10,0	5,0	10,0	154,2
IT10DB85	bdl.	154,2	5,0	417,9	5,0	5,0	10,0	99,5
IT09DB11	bdl.	240,0	10,0	15,0	5,0	bdl.	30,0	10,0
IT10DB86	bdl.	19,8	5,0	44,6	5,0	5,0	9,9	14,9
IT10DB87	bdl.	265,0	5,0	250,0	5,0	20,0	20,0	60,0
IT10DB88	10,0	69,9	5,0	64,9	5,0	10,0	20,0	5,0
IT10DB89	15,0	354,6	5,0	314,7	10,0	25,0	20,0	84,9
IT10DB90	14,9	503,0	5,0	209,2	5,0	5,0	10,0	49,8
IT08DB07	5,0	275,0	10,0	20,0	5,0	bdl.	40,0	bdl.
IT10DB91	bdl.	283,3	5,0	238,6	5,0	5,0	14,9	59,6
IT10DB92	bdl.	239,5	5,0	189,6	5,0	5,0	15,0	44,9
IT10DB93	5,0	215,0	5,0	195,0	5,0	5,0	15,0	45,0
IT08DB04	bdl.	230,0	10,0	10,0	5,0	bdl.	35,0	bdl.
IT10DB94	bdl.	189,8	5,0	249,8	10,0	5,0	25,0	59,9
IT10DB95	bdl.	254,5	5,0	154,7	10,0	bdl.	15,0	34,9

Pb (ppm) ICP-OES	Rb (ppm) ICP-OES	Sb (ppm) ICP-OES	Sc (ppm) ICP-OES	Sn (ppm) ICP-OES	Sr (ppm) ICP-OES	V (ppm) ICP-OES	W (ppm) ICP-OES	Zn (ppm) ICP-OES	Zr (ppm) ICP-OES
25,0	159,7	10,0	5,0	10,0	94,8	29,9	bdl.	29,9	79,8
14,9	124,5	10,0	5,0	bdl.	164,3	24,9	bdl.	44,8	209,2
19,9	bdl.	bdl.	5,0	9,9	139,2	49,7	bdl.	54,7	188,9
10,0	34,9	10,0	5,0	10,0	174,3	5,0	bdl.	5,0	49,8
50,0	15,0	bdl.	bdl.	20,0	215,0	5,0	bdl.	15,0	25,0
10,0	14,9	19,9	bdl.	10,0	1145,4	bdl.	bdl.	bdl.	24,9
10,0	bdl.	10,0	bdl.	10,0	69,9	5,0	bdl.	bdl.	189,6
19,9	34,9	10,0	bdl.	bdl.	99,7	5,0	bdl.	bdl.	99,7
9,9	bdl.	bdl.	5,0	5,0	451,8	5,0	bdl.	bdl.	19,9
bdl.	114,7	5,0	bdl.	bdl.	169,5	bdl.	bdl.	5,0	10,0
24,9	bdl.	10,0	5,0	15,0	413,8	bdl.	bdl.	5,0	29,9
20,0	bdl.	bdl.	5,0	bdl.	390,0	bdl.	bdl.	bdl.	25,0
20,0	110,0	10,0	5,0	5,0	320,0	5,0	bdl.	10,0	35,0
5,0	bdl.	bdl.	5,0	5,0	294,4	bdl.	bdl.	5,0	49,9
14,9	bdl.	bdl.	5,0	9,9	367,4	19,9	bdl.	9,9	34,8
bdl.	bdl.	9,9	5,0	5,0	1325,7	bdl.	bdl.	bdl.	9,9
9,9	19,9	9,9	9,9	bdl.	223,7	39,8	bdl.	29,8	59,6
10,0	169,3	10,0	10,0	bdl.	94,6	59,8	bdl.	84,7	119,5
24,9	164,0	14,9	9,9	5,0	218,7	84,5	bdl.	59,6	159,0
29,8	173,8	9,9	14,9	9,9	129,1	44,7	5,0	44,7	119,2
25,0	160,0	bdl.	10,0	10,0	195,0	45,0	bdl.	30,0	105,0
20,0	174,7	10,0	10,0	10,0	139,7	64,9	bdl.	39,9	104,8
20,0	185,0	10,0	5,0	5,0	130,0	25,0	bdl.	30,0	70,0
19,9	169,0	9,9	5,0	9,9	119,3	19,9	bdl.	14,9	64,6
15,0	90,0	15,0	5,0	bdl.	220,0	25,0	bdl.	25,0	85,0
bdl.	100,0	10,0	5,0	bdl.	170,0	5,0	bdl.	10,0	80,0
bdl.	104,6	10,0	5,0	bdl.	134,5	24,9	bdl.	24,9	79,7
25,0	89,9	10,0	5,0	bdl.	89,9	40,0	bdl.	50,0	99,9
bdl.	74,8	10,0	10,0	bdl.	204,4	44,9	bdl.	44,9	119,6
bdl.	19,9	bdl.	5,0	bdl.	194,0	14,9	bdl.	29,9	44,8
bdl.	39,8	bdl.	5,0	bdl.	179,1	5,0	bdl.	10,0	44,8
bdl.	40,0	15,0	5,0	bdl.	205,0	5,0	5,0	bdl.	35,0
14,9	9,9	bdl.	5,0	bdl.	2123,0	bdl.	bdl.	bdl.	19,8
bdl.	35,0	10,0	5,0	bdl.	335,0	10,0	bdl.	bdl.	60,0
bdl.	10,0	bdl.	5,0	bdl.	1458,5	bdl.	bdl.	5,0	40,0
10,0	50,0	10,0	15,0	bdl.	364,6	50,0	bdl.	20,0	104,9
bdl.	84,7	bdl.	5,0	bdl.	353,6	5,0	bdl.	10,0	44,8
20,0	60,0	15,0	5,0	bdl.	425,0	5,0	bdl.	0,0	30,0
bdl.	59,6	9,9	5,0	bdl.	238,6	bdl.	bdl.	9,9	29,8
bdl.	49,9	15,0	5,0	bdl.	289,4	5,0	bdl.	10,0	29,9
bdl.	35,0	20,0	5,0	bdl.	390,0	bdl.	bdl.	10,0	35,0
bdl.	35,0	5,0	5,0	bdl.	200,0	bdl.	5,0	bdl.	75,0
35,0	40,0	10,0	5,0	bdl.	179,8	bdl.	5,0	5,0	30,0
bdl.	64,9	10,0	5,0	bdl.	379,2	bdl.	bdl.	10,0	25,0

Sample	As	Ba	Co	Cr	Cu	La	Nd	Ni
	(ppm)	(ppm)	(ppm)	(ppm)	(ppm)	(ppm)	(ppm)	(ppm)
	ICP-OES	ICP-OES	ICP-OES	ICP-OES	ICP-OES	ICP-OES	ICP-OES	ICP-OES
IT10DB96	9,9	457,3	5,0	178,9	5,0	9,9	14,9	39,8
IT10DB97	bdl.	244,8	5,0	254,7	5,0	20,0	15,0	59,9
IT08DB03	bdl.	375,0	10,0	65,0	5,0	15,0	40,0	15,0
IT10DB98	5,0	352,5	5,0	238,3	5,0	9,9	19,9	59,6
IT09DB10	bdl.	355,0	10,0	75,0	5,0	10,0	30,0	15,0
IT10DB99	bdl.	285,0	5,0	310,0	5,0	5,0	15,0	95,0
IT10DB100	10,0	264,5	5,0	289,4	5,0	15,0	15,0	99,8
IT10DB101	bdl.	239,3	5,0	378,9	10,0	24,9	15,0	109,7
IT10DB102	bdl.	224,8	5,0	149,9	5,0	30,0	15,0	40,0
IT10DB103	bdl.	20,0	5,0	29,9	bdl.	5,0	15,0	5,0
IT10DB104	bdl.	24,8	5,0	39,7	5,0	bdl.	19,8	9,9
IT09DB29	20,0	90,0	5,0	30,0	bdl.	bdl.	40,0	5,0
Shell	bdl.	5,0	10,0	bdl.	bdl.	bdl.	bdl.	bdl.
Na_2CO_3	10,0	bdl.	10,0	bdl.	bdl.	bdl.	bdl.	5,0
PO09DB01	bdl.	95,0	10,0	5,0	bdl.	bdl.	15,0	bdl.
GR08DB01	bdl.	95,0	10,0	3555,0	bdl.	5,0	40,0	20,0
GR09DB02	5,0	30,0	5,0	30,0	5,0	bdl.	40,0	5,0
GR09DB03	10,0	1625,0	5,0	40,0	65,0	55,0	30,0	20,0
TU08DB01	5,0	375,0	5,0	25,0	5,0	bdl.	20,0	10,0
CY09DB01	5,0	110,0	35,0	1580,0	30,0	10,0	25,0	535,0
TN09DB01	5,0	30,0	10,0	10,0	5,0	bdl.	30,0	bdl.
LI09DB01	bdl.	100,0	5,0	15,0	5,0	bdl.	20,0	bdl.
LI09DB02	bdl.	45,0	5,0	10,0	bdl.	bdl.	20,0	bdl.
LI09DB03	20,0	60,0	5,0	10,0	5,0	bdl.	20,0	10,0
LI09DB04	bdl.	55,0	5,0	10,0	5,0	5,0	20,0	bdl.
LI09DB05	25,0	120,0	5,0	15,0	5,0	5,0	25,0	5,0
EG08DB01	bdl.	110,0	5,0	5,0	5,0	bdl.	25,0	bdl.
EG08DB02	bdl.	65,0	5,0	10,0	5,0	bdl.	20,0	bdl.
EG08DB03	bdl.	45,0	10,0	10,0	5,0	bdl.	25,0	bdl.
EG08DB04	bdl.	40,0	10,0	15,0	bdl.	bdl.	25,0	5,0
EG08DB05	bdl.	20,0	5,0	10,0	10,0	bdl.	45,0	bdl.
EG09DB06	bdl.	540,0	15,0	50,0	bdl.	45,0	45,0	15,0
EG09DB09	bdl.	145,0	5,0	20,0	5,0	10,0	25,0	5,0
EG09DB10	bdl.	155,0	10,0	25,0	10,0	bdl.	25,0	bdl.
EG09DB11	15,0	110,0	10,0	10,0	5,0	bdl.	25,0	10,0
EG09DB12	bdl.	275,0	5,0	35,0	10,0	bdl.	40,0	10,0
EG09DB13	15,0	95,0	5,0	10,0	5,0	bdl.	20,0	5,0
EG10DB14	bdl.	34,8	9,9	248,5	5,0	bdl.	9,9	64,6
EG10DB15	bdl.	39,9	5,0	144,6	34,9	5,0	15,0	34,9
EG10DB16	bdl.	25,0	5,0	170,0	15,0	5,0	15,0	25,0
IS09DB04	bdl.	10,0	5,0	10,0	5,0	bdl.	40,0	10,0
IS09DB05	10,0	20,0	5,0	10,0	5,0	bdl.	45,0	5,0
IS09DB06	bdl.	1085,0	5,0	55,0	10,0	bdl.	50,0	20,0
IS09DB07	bdl.	800,0	5,0	35,0	10,0	25,0	25,0	10,0

Pb (ppm) ICP-OES	Rb (ppm) ICP-OES	Sb (ppm) ICP-OES	Sc (ppm) ICP-OES	Sn (ppm) ICP-OES	Sr (ppm) ICP-OES	V (ppm) ICP-OES	W (ppm) ICP-OES	Zn (ppm) ICP-OES	Zr (ppm) ICP-OES
bdl.	9,9	9,9	5,0	bdl.	591,5	bdl.	bdl.	19,9	59,6
bdl.	30,0	bdl.	10,0	bdl.	399,6	25,0	bdl.	30,0	94,9
bdl.	45,0	bdl.	5,0	bdl.	520,0	20,0	bdl.	bdl.	75,0
bdl.	54,6	14,9	5,0	bdl.	297,9	14,9	bdl.	19,9	74,5
bdl.	55,0	bdl.	5,0	10,0	315,0	20,0	bdl.	bdl.	95,0
25,0	55,0	10,0	5,0	bdl.	315,0	15,0	5,0	20,0	50,0
10,0	64,9	15,0	10,0	bdl.	264,5	34,9	bdl.	44,9	74,9
10,0	49,9	10,0	10,0	bdl.	264,2	39,9	bdl.	44,9	94,7
bdl.	45,0	5,0	10,0	bdl.	159,8	45,0	bdl.	74,9	104,9
bdl.	bdl.	bdl.	5,0	5,0	154,7	15,0	bdl.	bdl.	20,0
bdl.	bdl.	9,9	5,0	bdl.	178,6	14,9	bdl.	bdl.	19,8
bdl.	15,0	25,0	5,0	bdl.	1285,0	bdl.	bdl.	bdl.	25,0
bdl.	nd.	bdl.	bdl.	bdl.	1568,4	bdl.	bdl.	bdl.	10,0
10,0	nd.	bdl.	bdl.	bdl.	bd.	bdl.	bdl.	bdl.	bdl.
bdl.	25,0	bdl.	bdl.	bdl.	90,0	bdl.	bdl.	bdl.	30,0
bdl.	bdl.	20,0	5,0	bdl.	255,0	40,0	bdl.	bdl.	375,0
bdl.	bdl.	20,0	5,0	bdl.	310,0	bdl.	bdl.	bdl.	15,0
35,0	125,0	20,0	10,0	bdl.	760,0	65,0	bdl.	bdl.	250,0
20,0	95,0	bdl.	5,0	5,0	200,0	15,0	bdl.	bdl.	60,0
bdl.	15,0	15,0	20,0	bdl.	160,0	100,0	bdl.	30,0	75,0
bdl.	bdl.	20,0	5,0	5,0	150,0	bdl.	bdl.	bdl.	25,0
bdl.	10,0	15,0	5,0	bdl.	25,0	5,0	bdl.	bdl.	285,0
bdl.	5,0	20,0	5,0	bdl.	20,0	5,0	bdl.	bdl.	90,0
bdl.	10,0	10,0	5,0	bdl.	20,0	5,0	bdl.	bdl.	130,0
bdl.	bdl.	40,0	5,0	bdl.	30,0	5,0	bdl.	bdl.	65,0
bdl.	20,0	20,0	5,0	10,0	85,0	20,0	bdl.	bdl.	105,0
bdl.	20,0	20,0	5,0	5,0	25,0	bdl.	bdl.	bdl.	165,0
bdl.	15,0	15,0	5,0	bdl.	15,0	bdl.	10,0	bdl.	40,0
5,0	bdl.	5,0	5,0	bdl.	15,0	bdl.	bdl.	130,0	35,0
10,0	bdl.	10,0	5,0	bdl.	15,0	5,0	5,0	bdl.	55,0
bdl.	bdl.	bdl.	5,0	bdl.	1840,0	bdl.	bdl.	65,0	20,0
15,0	65,0	15,0	5,0	bdl.	685,0	140,0	bdl.	40,0	370,0
75,0	350,0	30,0	5,0	bdl.	45,0	15,0	bdl.	bdl.	180,0
10,0	10,0	bdl.	5,0	bdl.	275,0	15,0	bdl.	bdl.	115,0
bdl.	bdl.	25,0	5,0	bdl.	155,0	5,0	bdl.	10,0	65,0
15,0	bdl.	15,0	5,0	5,0	445,0	20,0	bdl.	5,0	170,0
bdl.	5,0	bdl.	5,0	5,0	40,0	bdl.	bdl.	bdl.	160,0
14,9	nd.	bdl.	nd.	bdl.	44,7	bdl.	bdl.	bdl.	24,9
69,8	nd.	bdl.	nd.	5,0	19,9	5,0	bdl.	bdl.	79,8
15,0	nd.	bdl.	nd.	5,0	50,0	5,0	bdl.	bdl.	30,0
bdl.	bdl.	15,0	5,0	10,0	3045,0	bdl.	bdl.	bdl.	15,0
bdl.	bdl.	10,0	5,0	5,0	2460,0	bdl.	bdl.	bdl.	10,0
bdl.	15,0	bdl.	5,0	10,0	545,0	35,0	bdl.	10,0	70,0
bdl.	105,0	15,0	5,0	5,0	480,0	40,0	bdl.	5,0	245,0

Sample	As	Ba	Co	Cr	Cu	La	Nd	Ni
	(ppm)	(ppm)	(ppm)	(ppm)	(ppm)	(ppm)	(ppm)	(ppm)
	ICP-OES	ICP-OES	ICP-OES	ICP-OES	ICP-OES	ICP-OES	ICP-OES	ICP-OES
IS09DB08	bdl.	160,0	10,0	10,0	5,0	bdl.	25,0	10,0
OM10DB01	bdl.	295,0	10,0	1195,0	45,0	bdl.	bdl.	75,0
NE08DB01	bdl.	188,9	5,0	34,8	94,4	9,9	19,9	bdl.
UK08DB01	bdl.	149,7	5,0	189,6	20,0	5,0	29,9	34,9
UK08DB02	bdl.	134,6	5,0	383,8	15,0	5,0	24,9	79,8

Pb	Rb	Sb	Sc	Sn	Sr	V	W	Zn	Zr
(ppm)	(ppm)	(ppm)	(ppm)	(ppm)	(ppm)	(ppm)	(ppm)	(ppm)	(ppm)
ICP-OES	ICP-OES	ICP-OES	ICP-OES	ICP-OES	ICP-OES	ICP-OES	ICP-OES	ICP-OES	ICP-OES
bdl.	bdl.	bdl.	5,0	10,0	230,0	5,0	10,0	bdl.	95,0
40,0	nd.	bdl.	nd.	5,0	385,0	25,0	bdl.	475,0	200,0
54,7	nd.	bdl.	nd.	5,0	49,7	bdl.	bdl.	14,9	124,3
bdl.	nd.	bdl.	nd.	15,0	15,0	15,0	bdl.	5,0	25,0
19,9	nd.	bdl.	nd.	bdl.	19,9	15,0	bdl.	bdl.	24,9

Sample	Nd	Cl	TiO$_2$	MnO	Sc	V	Cr	Co
	(ppm)	(Wt.%)	(Wt.%)	(Wt.%)	(ppm)	(ppm)	(ppm)	(ppm)
	ICP-MS	INAA	INAA	INAA	INAA	INAA	INAA	INAA
Spain								
SP10DB46	9,2	0,15	0,31	0,05	3,90	32,9	22,6	4,52
SP10DB45	5,6	0,12	0,61	0,03	2,20	31,8	16,8	3,26
SP10DB43	5,6	0,34	0,16	0,01	1,17	12,8	8,2	1,54
SP10DB42	7,3							
SP10DB37	5,8							
SP10DB32	14,6							
SP10DB28	nd.							
SP10DB27	9,4							
SP10DB22	9,1	0,11	0,27	0,05	4,30	31,2	19,1	4,53
SP10DB20	7,0	0,17	0,16	0,02	2,86	23,5	14,7	3,07
SP08DB01	nd.							
SP10DB17	5,4							
SP10DB14	nd.							
SP10DB10	6,9							
SP10DB09	5,5							
SP10DB08	nd.							
SP10DB07	12,1							
SP10DB06	15,0							
SP08DB02	7,0	0,01	0,10	0,09	2,37	12,9	124,9	2,32
SP10DB05	6,6	0,01	0,34	0,02	2,24	24,9	277,0	1,81
SP08FH08	nd.							
SP08FH01	nd.							
SP08FH12	11,1							
SP08FH11	25,2							
SP08FH09	39,1							
SP09DB03	296,3							
SP10DB48	59,5							
FR09DB02	59,5							
FR09DB05	nd.							
FR09DB07	nd.							
FR09DB10	nd.							
FR09DB12	nd.							
FR09DB17	nd.							
FR09DB20	7,3							
IT09DB12	nd.							
IT09DB13	6,7							
IT09DB16b	nd.							
IT09DB18	nd.							
IT10DB30	13,4							
IT08DB05	nd.							
IT10DB37	38,7							

Ni	Cu	Zn	Ga	Se	Br	Rb	Sr	Zr	Ag
(ppm)	(ppm)	(ppm)	(ppm)	(ppm)	(ppm)	(ppm)	(ppm)	(ppm)	(ppm)
INAA	INAA	INAA	INAA	INAA	INAA	INAA	INAA	INAA	INAA
<41	<47	47	<20	<0.87	<6.4	35,3	84	67	<1.9
<37	<39	84	<17	<0.82	<6.1	25,5	82	170	<2.0
<26	<41	11,7	<17	<0.59	11,3	38,9	94	42	<1.2
21	<45	47	13	<0.91	<6.4	25,4	107	<180	<2.0
<36	<37	32	<16	<0.77	3,5	24,1	32	71	<1.8
	<35	<30	16	<26	<0.74	<9.3	36,1	192	<68
	<35	<38	15	<23	<0.79	<6.2	28,6	50	<220

Sample	Nd	Cl	TiO$_2$	MnO	Sc	V	Cr	Co
	(ppm)	(Wt.%)	(Wt.%)	(Wt.%)	(ppm)	(ppm)	(ppm)	(ppm)
	ICP-MS	INAA	INAA	INAA	INAA	INAA	INAA	INAA
IT10DB41	7,7							
IT10DB42	10,5	0,10	0,22	0,11	4,05	22,0	72,2	3,68
IT08DB06	nd.							
IT10DB44	10,7	0,14	0,08	0,03	1,17	14,8	16,2	1,08
IT09DB23	nd.							
IT09DB21	nd.							
IT09DB25	7,2							
IT09DB26	nd.							
IT09DB20	nd.							
IT09DB28	8,5							
IT08DB02	nd.							
IT10DB49	nd.							
IT10DB50	nd.							
IT10DB52	nd.							
IT10DB55	nd.							
IT10DB62	1,01	0,12	<0.01	0,00	0,04	<0.64	0,93	0,20
IT10DB63	0,95							

Ni	Cu	Zn	Ga	Se	Br	Rb	Sr	Zr	Ag
(ppm)	(ppm)	(ppm)	(ppm)	(ppm)	(ppm)	(ppm)	(ppm)	(ppm)	(ppm)
INAA	INAA	INAA	INAA	INAA	INAA	INAA	INAA	INAA	INAA
<42	<54	15	<31	<0.91	<10	30,7	196	190	0,8
<26	<35	5,3	<16	<0.59	5,4	24,2	315	40	<1.2
<56	<33	1,2	<15	<0.36	5,0	<1.2	1550	<16	<0.6

Sample	In	Sb	I	Cs	Ba	Ce	Nd	Eu	Tb	Dy
	(ppm)	(ppm)	(ppm)	(ppm)	(ppm)	(ppm)	(ppm)	(ppm)	(ppm)	(ppm)
	INAA	INAA	INAA	INAA	INAA	INAA	INAA	INAA	INAA	INAA
SP10DB46	<0.049	0,65	<2.0	1,11	180	<140	11,0	0,46	0,28	2,01
SP10DB45	<0.045	0,71	1,10	0,75	110	<130	11,1	0,42	0,26	1,75
SP10DB43	<0.039	0,44	<1.4	1,16	159	<120	4,8	0,27	0,11	0,81
SP10DB42										
SP10DB37										
SP10DB32										
SP10DB28										
SP10DB27										
SP10DB22	<0.054	1,38	<2.0	1,63	169	<140	9,0	0,40	0,25	2,04
SP10DB20	0,026	0,90	<1.4	4,07	95	<120	6,2	0,28	0,17	1,54
SP08DB01										
SP10DB17										
SP10DB14										
SP10DB10										
SP10DB09										
SP10DB08										
SP10DB06										
SP08DB02										
SP10DB04										
SP08FH06										
SP08FH08										
SP10DB47										
SP08FH04										
SP08FH03										
SP08FH11										
SP08FH10										
SP10DB48										
SP10DB49										
FR09DB02										
FR09DB04										
FR09DB08										
FR09DB10										
FR09DB13										
FR09DB16	<0.09	19,2	<2.7	0,88	150	<150	8,8	0,51	0,26	1,79
FR09DB17	<0.085	0,42	<3.1	0,48	186	<120	4,5	0,47	0,19	1,46
FR09DB18										
FR09DB19										
IT09DB12										
IT09DB13										
IT09DB17										
IT09DB09										
IT10DB30										
IT10DB31										

Dy (ppm) INAA	Yb (ppm) INAA	Lu (ppm) INAA	Hf (ppm) INAA	Ta (ppm) INAA	Hg (ppm) INAA	Th (ppm) INAA	$^{87}Sr/^{86}Sr$	2s	$^{143}Nd/^{144}Nd$	2s	ε_{Nd}
2,01	0,78	0,13	1,80	0,34	<0.20	2,30	0,71378	0,00007	0,512229	0,000070	-7,99
1,75	0,72	0,11	5,07	0,48	<0.26	3,37	0,71197	0,00010	0,512193	0,000042	-8,68
0,81	0,41	0,06	1,38	0,22	<0.24	1,86	0,71369	0,00007	0,512167	0,000037	-9,20
							0,70949	0,00009	0,512076	0,000051	-10,96
							0,71089	0,00011	0,512007	0,000073	-12,30
							0,71293	0,00011	0,512026	0,000032	-11,93
							0,71716	0,00010	0,512012	0,000050	-12,22
							0,71018	0,00008	0,512049	0,000090	-11,50
2,04	0,99	0,15	1,94	0,35	<0.19	2,74	0,71222	0,00009	0,512057	0,000037	-11,33
1,54	0,66	0,10	1,82	0,29	<0.16	2,15	0,71559	0,00018	0,512035	0,000050	-11,76
							0,70800	0,00007	nd.	nd.	nd.
							0,70937	0,00009	0,512141	0,000053	-9,69
							0,70885	0,00010	0,512109	0,000055	-10,32
							0,71022	0,00007	0,512100	0,000047	-10,50
							0,70984	0,00010	0,512126	0,000059	-9,99
							0,70965	0,00010	0,512090	0,000047	-10,70
							0,71003	0,00012	0,512044	0,000036	-11,58
							0,71050	0,00007	0,512018	0,000101	-12,10
							0,70928	0,00009	0,512075	0,000035	-10,98
							0,71032		0,512049		-11,49
							0,71520	0,00014	0,512133	0,000082	-9,84
							0,71253	0,00010	0,512104	0,000058	-10,42
							0,71298	0,00014	0,512065	0,000091	-11,18
							0,71584		0,512161		-9,30
							0,72264		0,512193		-8,68
							0,72232	0,00019	0,512166	0,000088	-9,20
							0,70928	0,00015	0,51215	0,00007	-9,52
							0,70931	0,00015	0,51213	0,00007	-9,98
							0,70928	0,00007	0,51207	0,00008	-11,06
							0,72901	0,00019	0,512124	0,000063	-10,03
							0,71013	0,00013	0,512190	0,000058	-8,74
							0,71177	0,00016	0,512212	0,000065	-8,31
							0,71388	0,00019	0,512170	0,000121	-9,14
1,79	0,73	0,12	2,45	0,47	<0.19	3,24	0,72358	0,00014	0,512002	0,000043	-12,40
1,46	0,81	0,12	2,81	0,48	<0.23	1,91	0,71315	0,00013	0,512118	0,000087	-10,14
							0,71587	0,00016	0,512203	0,000058	8,48
							0,71926	0,00026	0,512227	0,000113	-8,03
							0,70872	0,00014	0,512185	0,000090	-8,83
							0,71304	0,00015	0,512292	0,000070	-6,75
							0,71135	0,00015	0,512482	0,000049	-3,05
							0,71090	0,00007	0,512236	0,000070	-7,84
							0,71149	0,00016	0,512207	0,000073	-8,41
							0,71090	0,00017	0,512234	0,000086	-7,87

Sample	In	Sb	I	Cs	Ba	Ce	Nd	Eu	Tb	Dy
	(ppm)	(ppm)	(ppm)	(ppm)	(ppm)	(ppm)	(ppm)	(ppm)	(ppm)	(ppm)
	INAA	INAA	INAA	INAA	INAA	INAA	INAA	INAA	INAA	INAA
IT10DB34	<0.067	0,45	<2.8	4,26	136	<120	5,2	0,46	0,25	1,52
IT08DB01	<0.054	0,82	<2.0	0,97	129	<140	6,3	0,28	0,14	1,33
IT10DB39										
IT08DB06										
IT10DB44										
IT09DB22										
IT09DB23										
IT09DB21										
IT09DB25										
IT09DB20										
IT08DB02										
IT10DB47										
IT10DB50										
IT10DB52										
IT10DB56										
IT10DB59										
IT10DB62										
IT10DB63										
IT08DB08										
IT10DB68										
IT10DB71										
IT10DB75										
IT10DB80										
IT10DB84										
IT10DB85	<0.077	<0.42	<3.2	0,60	230	<150	14,7	0,60	0,35	2,60
IT09DB11										
IT10DB87	<0.041	0,22	<1.5	0,34	280	<96	11,6	0,55	0,25	1,73
IT10DB90										
IT08DB07										
IT10DB91										
IT10DB92										
IT08DB04										
IT10DB94										
IT10DB96										
IT09DB10										
IT10DB99										
IT10DB101										
IT10DB104										
IT10DB105										
IT10DB106										
IT10DB107										
IT10DB108										
Shell	<0.034	0,14	<1.2	0,04	15	<65	0,4	0,01	<0.022	<0.19
Na_2CO_3										

Dy (ppm) INAA	Yb (ppm) INAA	Lu (ppm) INAA	Hf (ppm) INAA	Ta (ppm) INAA	Hg (ppm) INAA	Th (ppm) INAA	$^{87}Sr/^{86}Sr$	2s	$^{143}Nd/^{144}Nd$	2s	ε_{Nd}
1,52	0,51	0,07	1,01	0,18	<0.16	1,55	0,71034	0,00013	0,512156	0,000028	-9,40
1,33	0,51	0,09	2,18	0,42	<0.17	2,85	0,71357	0,00002	0,512184	0,000009	-8,86
							0,71073	0,00009	0,512133	0,000088	-9,85
							0,71033	0,00007	0,512138	0,000037	-9,76
							0,70967	0,00010	0,512168	0,000106	-9,16
							0,70922	0,00001	0,512133	0,000010	-9,90
							0,71013	0,00014	0,512233	0,000054	-7,90
							0,70796	0,00002	0,512411	0,000006	-4,40
							0,70969	0,00001	0,512284	0,000009	-6,90
							0,70969	0,00001	0,512284	0,000009	-6,90
							0,70748	0,00001	0,512415	0,000008	-4,35
							0,71326	0,00013	0,512203	0,000070	-8,48
							0,71164	0,00011	0,512243	0,000146	-7,71
							0,71434	0,00009	0,512086	0,000051	-10,77
							0,72218	0,00012	0,512062	0,000035	-11,23
							0,70916	0,00013	0,512072	0,000110	-11,04
							0,70990	0,00020	0,511999	0,000102	-12,47
							0,70922	0,00010	0,511979	0,000114	-12,85
							0,70895	0,00007	0,512067	0,000116	-11,14
							0,70914	0,00005	0,512122	0,000051	-10,07
							0,70851	0,00015	0,512235	0,000092	-7,86
							0,71788	0,00008	0,512042	0,000057	-11,63
							0,71687	0,00013	0,512087	0,000045	-10,75
							0,71046	0,00016	0,512164	0,000116	-9,24
2,60	1,19	0,18	5,35	0,48	<0.70	5,45	0,71079	0,00054	0,512325	0,000078	-6,11
							0,71178	0,00007	0,512194	0,000011	-8,66
1,73	0,62	0,09	1,10	0,25	<0.13	2,06	0,70867	0,00012	0,512424	0,000076	-4,17
							0,70978	0,00014	0,512329	0,000089	-6,03
							0,71088	0,00007	0,512245	0,000036	-7,67
							0,71163	0,00010	0,512160	0,000033	-9,32
							0,71112	0,00010	0,512222	0,000071	-8,12
							0,71143	0,00001	0,512173	0,000011	-9,07
							0,71064	0,00015	0,512199	0,000066	-8,57
							0,70916	0,00015	0,512230	0,000092	-7,95
							0,71159	0,00007	0,512211	0,000059	-8,34
							0,71192	0,00016	0,512187	0,000083	-8,79
							0,71202	0,00011	0,512209	0,000063	-8,37
							0,70846	0,00013	0,512274	0,001189	-7,10
							0,72127	0,00011	0,512200	0,000030	-8,59
							0,70930	0,00012	0,512270	0,000040	-7,17
							0,71107	0,00012	0,512160	0,000060	-9,29
							0,72289	0,00012	0,512240	0,000070	-7,82
<0.19	0,02	<0.0073	0,03	0,02	<0.081	0,06	0,70915	0,00010	0,512297	0,001068	-6,66
							nd.	nd.	0,512626	0,003814	-0,23

Sample	In	Sb	I	Cs	Ba	Ce	Nd	Eu	Tb	Dy
	(ppm)	(ppm)	(ppm)	(ppm)	(ppm)	(ppm)	(ppm)	(ppm)	(ppm)	(ppm)
	INAA	INAA	INAA	INAA	INAA	INAA	INAA	INAA	INAA	INAA
PO09DB01										
GR08DB01										
GR13DB04										
TU08DB01										
TN09DB01										
TN10DB02										
LI09DB01										
LI09DB03										
LI09DB04										
LI09DB05										
EG08DB01										
EG08DB02										
EG08DB03										
EG08DB04										
EG08DB05										
EG09DB10										
EG09DB11										
EG09DB12										
EG09DB13										
EG10DB14										
EG10DB15										
EG10DB16										
EG10DB17										
EG13DB18										
IS08DB01										
IS08DB02										
IS08DB03										
IS09DB05										
IS13DB09										
IS13DB10										
IS13DB12										
IS13DB13										
BE08DB01										
NE08DB01										
UK08DB01										
UK08DB02										

Dy	Yb	Lu	Hf	Ta	Hg	Th	$^{87}Sr/^{86}Sr$	2s	$^{143}Nd/^{144}Nd$	2s	ε_{Nd}
(ppm)	(ppm)	(ppm)	(ppm)	(ppm)	(ppm)	(ppm)					
INAA	INAA	INAA	INAA	INAA	INAA	INAA					
							0,71275	0,00007	0,512204	0,000093	-8,47
							0,70910	0,00007	0,512304	0,000085	-6,51
							0,70859	0,00012	0,51242	0,00008	-4,21
							0,72091	0,00007	0,512259	0,000117	-7,40
							0,70833	0,00007	0,51207	0,00004	-11,00
							0,71329	0,00014	0,51199	0,00004	-12,57
							0,71821	0,00014	0,51226	0,00006	-7,39
							0,71451	0,00009	0,51204	0,00005	-11,67
							0,71651	0,00011	0,51191	0,00004	-14,24
							0,71410	0,00016	0,51200	0,00008	-12,54
							0,72108	0,00002	0,51211	0,00001	-10,36
							0,71950	0,00002	0,51212	0,00001	-10,03
							0,71525	0,00002	0,51210	0,00001	-10,46
							0,71055	0,00001	0,51212	0,00001	-10,18
							0,70915	0,00007	0,51243	0,00001	-3,99
							0,70811	0,00013	0,51233	0,00010	-5,99
							0,70841		0,51232		-6,15
							0,70810	0,00010	0,51235	0,00004	-5,69
							0,71142	0,00012	0,51230	0,00005	-6,65
							0,70921	0,00011	0,51219	0,00003	-8,67
							0,71148	0,00008	0,51230	0,00006	-6,52
							0,70831	0,00011	0,51216	0,00004	-9,32
							0,71292	0,00010	0,51239	0,00006	-4,85
							0,70826	0,00015	0,51239	0,00007	-4,75
							0,71159	0,00001	0,51239	0,00001	-4,80
							0,70920	0,00002	0,51239	0,00001	-4,80
							0,70925	0,00003	0,51259	0,00001	-1,00
							0,70919		0,51256		-1,60
							0,70927	0,00014	0,51246	0,00008	-3,38
							0,70929	0,00018	0,51241	0,00019	-4,45
							0,70912	0,00010	0,51244	0,00008	-3,92
							0,70947	0,00019	0,51228	0,00010	-6,96
							0,71487	0,00002	0,51199	0,00001	-12,70
							0,72082	0,00001	0,51214	0,00001	-9,71
							0,73620	0,00007	0,51205	0,00003	-11,56
							0,73732	0,00007	0,51204	0,00004	-11,66

Calculated glass compositions after raising the Na$_2$O levels of the sands to 16.63%, the average Na$_2$O content of Roman natron glass (Foster and Jackson, 2009)

Bold: values within compositional ranges (two times standard deviation; see Table 2.1). Italic: values within three times the standard deviation. All results are in wt%.

	SP46	SP45	SP44	SP43	SP42	SP41	SP40
SiO$_2$	*76,29*	78,27	80,70	77,92	**69,09**	58,17	59,36
Al$_2$O$_3$	**3,14**	*1,59*	1,10	**1,92**	1,10	1,10	0,29
Fe$_2$O$_3$(t)	**1,18**	**0,87**	**0,41**	**0,43**	**0,71**	**0,50**	**0,37**
MgO	**0,32**	**0,14**	**0,06**	**0,13**	**0,61**	**0,55**	**0,31**
MnO	**0,03**	**0,01**	**0,00**	**0,01**	**0,04**	**0,02**	**0,02**
CaO	1,36	1,78	0,66	2,08	11,29	22,48	22,86
Na$_2$O	**16,63**	**16,63**	**16,63**	**16,63**	**16,63**	**16,63**	**16,63**
K$_2$O	**0,87**	**0,58**	**0,39**	**0,81**	**0,42**	**0,40**	*0,09*
TiO$_2$	**0,15**	**0,08**	**0,04**	**0,05**	**0,06**	**0,09**	**0,01**
P$_2$O$_5$	**0,03**	**0,04**	*0,02*	**0,03**	**0,04**	**0,06**	**0,05**
Total	100,00	100,00	100,00	100,00	100,00	100,00	100,00
Added Na$_2$O	16,08	16,36	16,47	16,12	16,41	16,21	16,42
Sand-derived Na$_2$O	0,55	0,27	0,16	0,51	0,22	0,42	0,21

	SP30	SP29	SP28	SP27	SP26	SP25	SP24
SiO$_2$	53,30	*63,94*	**72,83**	**70,43**	**71,19**	*64,03*	49,16
Al$_2$O$_3$	17,60	5,80	4,49	**3,30**	*3,71*	**2,66**	8,83
Fe$_2$O$_3$(t)	4,14	2,82	2,45	*2,05*	2,23	*1,82*	6,60
MgO	2,44	3,27	**0,69**	*1,45*	**1,16**	1,62	5,57
MnO	**0,10**	**0,05**	**0,03**	**0,05**	**0,04**	**0,04**	**0,12**
CaO	4,75	**6,23**	1,90	**5,36**	*4,06*	12,08	*10,91*
Na$_2$O	**16,63**	**16,63**	**16,63**	**16,63**	**16,63**	**16,63**	**16,63**
K$_2$O	**0,55**	**0,92**	**0,66**	**0,53**	**0,62**	**0,45**	1,57
TiO$_2$	*0,44*	**0,27**	**0,24**	**0,16**	**0,31**	0,64	*0,54*
P$_2$O$_5$	**0,06**	**0,08**	**0,07**	**0,05**	**0,05**	**0,04**	**0,07**
Total	100,00	100,00	100,00	100,00	100,00	100,00	100,00
Added Na$_2$O	16,24	15,93	16,02	16,15	16,13	16,20	15,26
Sand-derived Na$_2$O	0,39	0,70	0,61	0,48	0,50	0,43	1,37

SP39	SP38	SP37	SP36	SP35	SP34	SP33	SP32	SP31
59,55	**69,32**	78,12	**65,65**	59,65	61,20	60,42	*64,21*	60,56
0,27	0,34	1,37	7,54	7,00	5,04	6,56	5,95	10,73
0,38	**0,56**	**0,72**	3,88	3,24	3,03	3,80	3,41	4,93
0,31	**0,21**	**0,68**	2,92	6,86	6,14	4,44	3,01	*1,23*
0,02	**0,01**	**0,01**	**0,13**	**0,04**	**0,04**	**0,09**	**0,06**	**0,12**
22,69	12,78	2,20	2,32	**5,33**	**6,98**	**6,64**	**5,46**	*4,04*
16,63	**16,63**	**16,63**	**16,63**	**16,63**	**16,63**	**16,63**	**16,63**	**16,63**
0,09	*0,12*	*0,18*	**0,65**	**0,96**	**0,70**	**1,05**	**0,93**	**1,22**
0,01	**0,02**	**0,07**	**0,23**	**0,24**	**0,19**	**0,32**	**0,26**	*0,44*
0,05	**0,02**	**0,02**	**0,05**	**0,05**	**0,04**	**0,07**	**0,08**	**0,09**
100,00	100,00	100,00	100,00	100,00	100,00	100,00	100,00	100,00
16,42	16,51	16,36	16,15	16,06	16,03	16,07	16,08	16,03
0,21	0,12	0,27	0,48	0,57	0,60	0,56	0,55	0,60

SP23	SP22	SP21	SP20	SP19	SP18	SP01	SP17	SP16	SP15
60,75	**72,17**	61,25	77,71	61,75	32,17	38,63	61,62	49,38	29,58
5,47	*3,41*	6,28	**2,36**	5,56	**1,85**	**2,47**	0,78	*1,54*	0,49
6,51	*1,71*	6,00	**1,17**	4,43	2,05	2,26	**0,23**	**0,75**	**0,50**
2,18	**0,96**	1,92	**0,45**	2,30	11,53	6,87	1,52	4,71	*1,40*
0,26	**0,03**	**0,21**	**0,01**	**0,18**	**0,06**	**0,06**	**0,01**	**0,02**	**0,01**
7,45	*4,33*	**6,78**	1,05	**7,97**	35,12	32,35	18,75	26,13	51,15
16,63	**16,63**	**16,63**	**16,63**	**16,63**	**16,63**	**16,63**	**16,63**	**16,63**	**16,63**
0,43	**0,54**	**0,54**	**0,44**	**0,85**	**0,40**	**0,46**	**0,40**	**0,71**	*0,17*
0,27	**0,17**	**0,33**	**0,13**	**0,26**	**0,14**	**0,23**	**0,04**	**0,09**	**0,03**
0,06	**0,04**	**0,06**	**0,04**	**0,07**	**0,06**	**0,04**	**0,02**	**0,05**	**0,05**
100,00	100,00	100,00	100,00	100,00	100,00	100,00	100,00	100,00	100,00
16,13	16,12	15,98	16,33	15,88	16,32	16,18	16,54	16,47	16,44
0,50	0,51	0,65	0,30	0,75	0,31	0,45	0,09	0,16	0,19

	SP14	SP13	SP12	SP11	SP10	SP09	SP08
SiO_2	*62,54*	57,37	25,04	60,36	61,52	**69,57**	62,72
Al_2O_3	0,70	1,34	0,84	0,94	**1,93**	0,92	**1,93**
$Fe_2O_3(t)$	**0,29**	**0,75**	**0,81**	**0,56**	**0,72**	**0,39**	**0,92**
MgO	**0,51**	**0,94**	3,69	*1,39*	**0,74**	**0,37**	**0,67**
MnO	**0,01**	**0,02**	**0,03**	**0,02**	**0,02**	**0,01**	**0,02**
CaO	19,01	22,48	52,48	19,61	17,52	11,61	16,47
Na_2O	**16,63**	**16,63**	**16,63**	**16,63**	**16,63**	**16,63**	**16,63**
K_2O	*0,25*	**0,33**	**0,39**	**0,43**	**0,80**	**0,42**	**0,50**
TiO_2	**0,04**	**0,10**	**0,03**	**0,04**	**0,08**	**0,05**	**0,09**
P_2O_5	**0,03**	**0,03**	**0,07**	**0,03**	**0,04**	**0,03**	**0,04**
Total	100,00	100,00	100,00	100,00	100,00	100,00	100,00
Added Na_2O	16,55	16,47	16,29	16,48	16,44	16,51	16,40
Sand-derived Na_2O	0,08	0,16	0,34	0,15	0,19	0,12	0,23

	FH04	FH03	FH01	FH02	FH12	FH11	FH10
SiO_2	**66,29**	68,49	**71,70**	**71,43**	**74,33**	74,47	71,56
Al_2O_3	4,60	9,75	7,72	7,81	5,72	5,67	7,66
$Fe_2O_3(t)$	**1,45**	**0,68**	**0,68**	**0,69**	**0,48**	**0,51**	**0,48**
MgO	**1,12**	**0,19**	**0,14**	**0,15**	**0,05**	**0,07**	**0,12**
MnO	**0,02**	**0,01**	**0,02**	**0,01**	**0,01**	**0,01**	**0,01**
CaO	**8,23**	1,49	0,67	0,48	0,28	0,29	0,58
Na_2O	**16,63**	**16,63**	**16,63**	**16,63**	**16,63**	**16,63**	**16,63**
K_2O	1,50	2,68	2,37	2,72	2,46	2,31	2,91
TiO_2	**0,12**	**0,04**	**0,04**	**0,05**	**0,02**	**0,02**	**0,02**
P_2O_5	**0,05**	**0,03**	**0,02**	**0,03**	**0,02**	**0,03**	**0,03**
Total	100,00	100,00	100,00	100,00	100,00	100,00	100,00
Added Na_2O	15,82	14,42	14,82	14,97	15,46	15,43	14,91
Sand-derived Na_2O	0,81	2,21	1,81	1,66	1,17	1,20	1,72

SP07	SP06	SP02	SP05	SP04	FH05	FH06	FH07	FH08	SP47
62,59	58,79	58,62	58,32	50,94	43,13	60,78	52,71	**67,48**	**68,09**
1,73	**3,14**	**3,01**	3,84	**2,30**	4,12	*3,52*	*3,62*	6,93	4,64
0,92	**0,95**	**1,41**	*1,89*	**1,14**	2,97	**1,48**	2,06	**1,18**	**0,78**
0,53	**0,72**	**0,84**	**0,97**	**0,63**	1,70	**0,67**	**1,11**	**0,69**	**0,69**
0,02	**0,02**	**0,03**	**0,04**	**0,01**	**0,06**	**0,03**	**0,04**	**0,02**	**0,01**
16,94	18,82	18,39	17,18	27,81	30,38	15,71	22,72	*4,34*	**7,64**
16,63	**16,63**	**16,63**	**16,63**	**16,63**	**16,63**	**16,63**	**16,63**	**16,63**	**16,63**
0,51	**0,75**	**0,63**	**0,61**	**0,44**	**0,64**	**1,00**	**0,80**	*2,55*	*1,41*
0,11	**0,13**	**0,38**	*0,45*	**0,06**	**0,29**	**0,14**	**0,24**	**0,14**	**0,06**
0,04	**0,04**	**0,05**	**0,06**	**0,04**	**0,08**	**0,05**	**0,07**	**0,04**	**0,04**
100,00	100,00	100,00	100,00	100,00	100,00	100,00	100,00	100,00	100,00
16,42	16,23	16,28	16,27	16,29	16,27	16,05	15,96	15,50	15,74
0,21	0,40	0,35	0,36	0,34	0,36	0,58	0,67	1,13	0,89

FH09	SP03	FR03	FR04	FR05	FR06	FR07	FR08	FR09	FR10
69,81	*63,11*	**68,57**	**70,11**	54,37	63,15	59,43	61,55	53,98	**67,39**
8,47	9,70	8,64	7,62	4,99	6,36	6,67	4,54	4,03	4,89
1,23	4,29	*1,89*	*1,61*	**1,43**	*1,91*	*1,92*	**1,12**	**0,89**	**0,97**
0,28	1,51	**0,63**	**0,44**	1,67	1,59	2,36	**0,80**	**0,58**	**0,50**
0,01	**0,04**	**0,02**	**0,02**	**0,06**	**0,04**	**0,04**	**0,04**	**0,03**	**0,03**
1,41	2,16	0,44	0,82	19,18	**7,84**	*10,24*	13,74	22,32	**7,92**
16,63	**16,63**	**16,63**	**16,63**	**16,63**	**16,63**	**16,63**	**16,63**	**16,63**	**16,63**
2,05	1,99	2,89	2,55	*1,31*	2,18	2,38	*1,24*	*1,35*	*1,41*
0,09	*0,45*	**0,22**	**0,14**	**0,27**	**0,23**	**0,25**	**0,26**	**0,14**	**0,20**
0,03	**0,12**	**0,07**	**0,04**	**0,08**	**0,08**	**0,09**	**0,09**	**0,06**	**0,06**
100,00	100,00	100,00	100,00	100,00	100,00	100,00	100,00	100,00	100,00
13,88	15,25	15,33	15,48	15,41	15,72	15,66	15,44	15,48	15,54
2,75	1,38	1,30	1,15	1,22	0,91	0,97	1,19	1,15	1,09

	FR11	FR12	FR13	FR14	FR16	FR17	FR18
SiO_2	50,35	16,03	76,99	35,82	76,80	**73,09**	**64,54**
Al_2O_3	5,78	0,55	*1,56*	0,68	3,59	6,61	5,94
$Fe_2O_3(t)$	4,17	**0,40**	**0,14**	**0,21**	**1,41**	**1,29**	**0,77**
MgO	2,31	2,02	**0,26**	**0,74**	**0,23**	**0,40**	**0,70**
MnO	**0,12**	**0,02**	**0,00**	**0,00**	**0,01**	**0,03**	**0,02**
CaO	17,68	63,84	3,26	45,53	0,36	0,89	**9,20**
Na_2O	**16,63**	**16,63**	**16,63**	**16,63**	**16,63**	**16,63**	**16,63**
K_2O	**0,94**	*0,16*	**1,11**	**0,33**	**0,69**	**0,66**	2,05
TiO_2	1,69	**0,05**	**0,04**	**0,04**	**0,25**	**0,37**	**0,11**
P_2O_5	0,34	0,29	*0,01*	*0,02*	**0,03**	**0,03**	**0,04**
Total	100,00	100,00	100,00	100,00	100,00	100,00	100,00
Added Na_2O	15,77	15,98	16,00	16,43	15,69	14,67	15,37
Sand-derived Na_2O	0,86	0,65	0,63	0,20	0,94	1,96	1,26

	IT18	IT19	IT09	IT30	IT31	IT32	IT33
SiO_2	47,52	60,23	**64,78**	**67,00**	**65,29**	33,90	50,14
Al_2O_3	10,45	7,49	6,00	6,17	5,99	6,64	5,81
$Fe_2O_3(t)$	5,69	2,77	*2,03*	*1,73*	*1,82*	3,59	5,03
MgO	12,72	3,13	**0,91**	**0,80**	**0,86**	7,92	9,99
MnO	**0,10**	**0,06**	**0,09**	**0,07**	**0,07**	**0,19**	**0,13**
CaO	**5,91**	**7,85**	**7,77**	**5,72**	**7,61**	30,61	11,46
Na_2O	**16,63**	**16,63**	**16,63**	**16,63**	**16,63**	**16,63**	**16,63**
K_2O	**0,61**	1,60	*1,41*	1,69	1,49	**0,27**	**0,50**
TiO_2	**0,33**	**0,19**	**0,33**	**0,14**	**0,18**	**0,21**	**0,26**
P_2O_5	**0,03**	**0,05**	**0,04**	**0,05**	**0,05**	**0,04**	**0,04**
Total	100,00	100,00	100,00	100,00	100,00	100,00	100,00
Added Na_2O	14,70	15,07	15,16	15,03	15,06	14,85	15,48
Sand-derived Na_2O	1,93	1,56	1,47	1,60	1,57	1,78	1,15

FR19	FR20	FR01	IT12	IT13	IT14	IT15	IT16a	IT16b	IT17
69,31	51,54	*57,54*	*64,36*	**69,49**	**68,05**	**66,09**	46,92	47,50	57,86
4,79	5,70	5,01	**2,66**	3,78	7,26	6,76	7,83	5,98	6,71
0,50	*1,61*	**0,54**	**1,00**	**1,17**	2,52	2,80	11,20	6,56	5,50
0,49	**0,93**	**1,13**	**0,37**	1,92	1,62	3,35	11,54	18,63	10,21
0,01	**0,03**	**0,01**	**0,04**	**0,03**	**0,04**	**0,05**	**0,13**	**0,12**	**0,07**
6,31	21,70	16,89	14,12	**5,94**	1,88	2,93	*4,04*	3,46	1,87
16,63	**16,63**	**16,63**	**16,63**	**16,63**	**16,63**	**16,63**	**16,63**	**16,63**	**16,63**
1,87	1,56	2,10	**0,71**	**0,87**	*1,38*	**0,92**	**0,82**	**0,65**	**0,76**
0,07	**0,19**	**0,08**	**0,10**	**0,14**	*0,52*	*0,42*	0,84	*0,42*	**0,36**
0,03	**0,12**	**0,08**	**0,02**	**0,03**	**0,09**	**0,05**	**0,05**	**0,04**	**0,04**
100,00	100,00	100,00	100,00	100,00	100,00	100,00	100,00	100,00	100,00
15,69	15,46	15,58	15,94	15,71	14,88	15,05	15,52	15,85	14,93
0,94	1,17	1,05	0,69	0,92	1,75	1,58	1,11	0,78	1,70

IT05	IT34	IT01	IT35	IT36	IT37	IT38	IT39	IT40	IT41
63,31	**71,42**	78,08	54,97	52,27	47,11	52,31	49,25	46,19	44,62
8,62	**2,55**	**2,63**	4,61	4,73	4,77	8,67	11,34	8,84	5,51
2,83	**0,89**	**0,81**	2,95	4,19	4,09	2,90	3,86	4,36	7,35
1,52	**0,31**	**0,19**	*1,33*	1,47	*1,24*	*1,46*	3,11	5,35	7,35
0,04	**0,07**	**0,02**	**0,22**	**0,17**	**0,27**	**0,18**	**0,12**	**0,11**	**0,13**
4,11	**7,18**	0,64	17,91	19,36	24,72	14,08	*10,68*	14,65	16,76
16,63	**16,63**	**16,63**	**16,63**	**16,63**	**16,63**	**16,63**	**16,63**	**16,63**	**16,63**
2,54	**0,86**	**0,70**	**0,96**	**0,87**	**0,97**	3,48	4,55	3,35	**0,73**
0,35	**0,07**	**0,28**	**0,36**	**0,24**	**0,15**	**0,23**	**0,37**	*0,42*	0,81
0,05	*0,02*	*0,01*	**0,05**	**0,06**	**0,05**	**0,06**	**0,10**	**0,10**	**0,11**
100,00	100,00	100,00	100,00	100,00	100,00	100,00	100,00	100,00	100,00
14,95	16,00	16,17	15,66	15,88	16,02	15,46	15,35	15,77	16,04
1,68	0,63	0,46	0,97	0,75	0,61	1,17	1,28	0,86	0,59

	IT42	IT06	IT43	IT44	IT45	IT23	IT24
SiO_2	42,09	43,60	56,27	60,68	48,66	49,71	55,01
Al_2O_3	4,93	5,07	5,84	**2,08**	5,21	5,07	9,94
$Fe_2O_3(t)$	6,70	7,70	2,28	**1,08**	3,49	2,95	*2,01*
MgO	9,17	6,90	1,52	*1,38*	4,37	3,65	**1,13**
MnO	**0,13**	**0,16**	**0,11**	**0,04**	**0,09**	**0,11**	**0,11**
CaO	19,00	18,35	15,39	16,94	19,53	20,05	11,51
Na_2O	**16,63**	**16,63**	**16,63**	**16,63**	**16,63**	**16,63**	**16,63**
K_2O	**0,57**	**0,64**	1,58	**1,08**	1,56	*1,32*	3,38
TiO_2	0,67	0,85	**0,26**	**0,05**	**0,38**	*0,41*	**0,21**
P_2O_5	**0,11**	**0,11**	**0,11**	**0,04**	**0,09**	**0,09**	**0,08**
Total	100,00	100,00	100,00	100,00	100,00	100,00	100,00
Added Na_2O	16,22	16,12	15,35	16,15	15,65	15,92	14,88
Sand-derived Na_2O	0,41	0,51	1,28	0,48	0,98	0,71	1,75

	IT51	IT52	IT53	IT54	IT55	IT56	IT57
SiO_2	*62,25*	**68,92**	**68,67**	**68,41**	**68,66**	**68,90**	*62,24*
Al_2O_3	11,42	8,81	9,49	9,27	8,28	7,60	*3,45*
$Fe_2O_3(t)$	3,80	**0,93**	**0,98**	**1,37**	**0,86**	2,58	3,24
MgO	1,73	**0,32**	**0,32**	**0,46**	**0,31**	0,84	**1,05**
MnO	**0,05**	**0,01**	**0,02**	**0,02**	**0,02**	**0,04**	**0,15**
CaO	1,34	1,71	1,35	1,40	2,53	1,33	12,50
Na_2O	**16,63**	**16,63**	**16,63**	**16,63**	**16,63**	**16,63**	**16,63**
K_2O	2,23	2,50	2,36	2,23	2,55	1,79	**0,40**
TiO_2	**0,46**	**0,11**	**0,12**	**0,15**	**0,09**	**0,22**	**0,28**
P_2O_5	**0,10**	**0,06**	**0,06**	**0,07**	**0,05**	**0,07**	**0,06**
Total	100,00	100,00	100,00	100,00	100,00	100,00	100,00
Added Na_2O	14,89	14,81	14,20	14,27	14,49	14,96	15,99
Sand-derived Na_2O	1,74	1,82	2,43	2,36	2,14	1,67	0,64

IT25	IT26	IT27	IT28	IT02	IT46	IT47	IT48	IT49	IT50
56,17	56,04	55,58	55,14	30,09	42,96	**70,85**	44,82	40,02	53,57
6,42	6,17	7,55	6,16	7,50	*3,44*	5,56	3,60	*1,45*	13,26
2,21	2,64	*1,74*	**1,37**	4,73	2,96	*1,71*	3,29	**1,42**	6,30
2,19	1,58	**0,98**	**1,10**	8,34	5,48	**0,60**	2,07	14,17	2,93
0,10	**0,13**	**0,13**	**0,14**	**0,09**	**0,14**	**0,04**	**0,06**	**0,06**	**0,10**
13,89	14,41	14,35	17,17	29,50	27,25	2,65	28,77	25,95	**5,28**
16,63	**16,63**	**16,63**	**16,63**	**16,63**	**16,63**	**16,63**	**16,63**	**16,63**	**16,63**
2,08	1,97	2,78	2,09	2,38	**0,76**	1,79	**0,51**	*0,18*	**1,09**
0,23	**0,33**	**0,17**	**0,13**	*0,53*	**0,23**	**0,13**	**0,18**	**0,07**	0,74
0,08	**0,10**	**0,07**	**0,07**	*0,22*	0,14	**0,04**	**0,06**	**0,04**	**0,10**
100,00	100,00	100,00	100,00	100,00	100,00	100,00	100,00	100,00	100,00
15,63	15,59	15,32	15,49	15,73	16,07	15,49	16,09	16,27	14,69
1,00	1,04	1,31	1,14	0,90	0,56	1,14	0,54	0,36	1,94

IT58	IT59	IT60	IT61	IT62	IT63	IT64	IT65	IT66	IT08
64,82	**72,45**	45,13	29,30	77,37	77,47	50,72	43,83	52,44	57,40
4,06	0,65	0,74	0,23	0,45	0,34	0,58	0,25	0,81	0,97
5,37	**1,04**	**1,46**	**0,33**	**0,87**	**0,48**	**1,15**	**0,43**	**1,50**	**1,41**
1,27	**0,78**	5,44	4,52	**0,30**	**0,11**	**0,50**	7,29	**0,59**	**0,43**
0,08	**0,03**	**0,02**	**0,02**	**0,02**	**0,02**	**0,06**	**0,02**	**0,08**	**0,08**
6,93	**8,22**	30,26	48,81	*4,12*	*4,74*	30,04	**31,41**	27,62	22,71
16,63	**16,63**	**16,63**	**16,63**	**16,63**	**16,63**	**16,63**	**16,63**	**16,63**	**16,63**
0,47	*0,12*	*0,14*	*0,05*	*0,14*	*0,13*	*0,17*	*0,08*	*0,23*	**0,28**
0,28	**0,05**	**0,05**	**0,02**	**0,07**	**0,06**	**0,03**	**0,02**	**0,03**	**0,04**
0,10	**0,03**	**0,12**	**0,08**	**0,03**	**0,02**	**0,13**	**0,05**	**0,07**	*0,05*
100,00	100,00	100,00	100,00	100,00	100,00	100,00	100,00	100,00	100,00
16,21	16,43	16,50	16,24	16,37	15,97	16,40	16,42	16,35	16,38
0,42	0,20	0,13	0,39	0,26	0,66	0,23	0,21	0,28	0,25

	IT67	IT68	IT69	IT70	IT71	IT72	IT73	IT74
SiO_2	*63,57*	*63,80*	59,02	2,40	**67,21**	*64,21*	60,78	*64,09*
Al_2O_3	*1,52*	1,01	1,13	0,63	1,27	9,45	11,63	9,97
$Fe_2O_3(t)$	**1,36**	**1,21**	**1,35**	**0,66**	3,56	5,20	4,77	4,07
MgO	**0,36**	**0,40**	**0,54**	2,21	1,47	*1,34*	1,69	**0,83**
MnO	**0,08**	**0,08**	**0,10**	**0,03**	**0,10**	**0,06**	**0,10**	**0,29**
CaO	15,94	16,44	20,76	77,20	**9,25**	0,92	1,98	1,78
Na_2O	**16,63**	**16,63**	**16,63**	**16,63**	**16,63**	**16,63**	**16,63**	**16,63**
K_2O	**0,45**	**0,31**	**0,33**	*0,11*	*0,16*	1,75	1,70	1,89
TiO_2	**0,05**	**0,05**	**0,08**	**0,05**	**0,30**	**0,35**	0,61	**0,33**
P_2O_5	**0,04**	**0,08**	**0,06**	**0,09**	**0,05**	**0,10**	**0,12**	**0,13**
Total	100,00	100,00	100,00	100,00	100,00	100,00	100,00	100,00
Added Na_2O	16,27	16,18	16,33	16,47	16,38	15,44	14,59	14,99
Sand-derived Na_2O	0,36	0,45	0,30	0,16	0,25	1,19	2,04	1,64

	IT85	IT11	IT86	IT87	IT88	IT89	IT90	IT07
SiO_2	**69,57**	**66,14**	9,04	**72,61**	22,07	57,42	*62,14*	51,44
Al_2O_3	**2,43**	3,73	0,48	1,14	1,14	4,20	5,59	4,52
$Fe_2O_3(t)$	**1,32**	**0,92**	**0,40**	**0,49**	**0,55**	2,58	**1,21**	**1,00**
MgO	**0,53**	**0,32**	3,71	**0,28**	2,31	2,20	**0,51**	**0,65**
MnO	**0,08**	**0,09**	**0,04**	**0,02**	**0,05**	**0,13**	**0,12**	**0,20**
CaO	**8,49**	*10,66*	69,45	**8,13**	56,58	15,22	11,26	23,53
Na_2O	**16,63**	**16,63**	**16,63**	**16,63**	**16,63**	**16,63**	**16,63**	**16,63**
K_2O	**0,81**	*1,45*	*0,16*	**0,63**	**0,48**	1,19	2,38	1,94
TiO_2	**0,10**	**0,06**	**0,03**	**0,04**	**0,05**	**0,29**	**0,10**	**0,07**
P_2O_5	**0,04**	*0,02*	**0,06**	**0,02**	**0,14**	**0,15**	**0,07**	**0,03**
Total	100,00	100,00	100,00	100,00	100,00	100,00	100,00	100,00
Added Na_2O	16,05	15,83	16,15	16,32	15,90	15,99	15,36	15,68
Sand-derived Na_2O	0,58	0,80	0,48	0,31	0,73	0,64	1,27	0,95

IT75	IT76	IT77	IT78	IT79	IT80	IT81	IT82	IT83	IT84
64,48	*64,21*	**65,31**	**66,75**	*63,95*	**67,14**	**67,30**	**66,27**	*64,08*	**64,60**
11,66	10,94	11,74	10,68	12,05	10,09	9,42	9,68	10,92	**3,21**
2,62	3,39	*1,88*	*1,65*	*2,05*	**1,01**	*1,75*	2,88	2,90	2,67
0,93	*1,18*	0,63	0,53	0,87	0,30	0,54	0,95	1,16	3,44
0,06	**0,05**	**0,03**	**0,03**	**0,03**	**0,02**	**0,03**	**0,04**	**0,05**	**0,08**
1,34	1,00	0,79	0,84	1,71	1,94	1,44	1,03	1,99	**8,66**
16,63	**16,63**	**16,63**	**16,63**	**16,63**	**16,63**	**16,63**	**16,63**	**16,63**	**16,63**
1,87	2,07	2,75	2,66	2,37	2,69	2,62	2,16	1,72	**0,56**
0,32	*0,43*	**0,19**	**0,17**	**0,27**	**0,10**	**0,20**	**0,27**	*0,46*	**0,12**
0,10	**0,10**	**0,06**	**0,07**	**0,07**	**0,08**	**0,07**	**0,10**	**0,08**	**0,04**
100,00	100,00	100,00	100,00	100,00	100,00	100,00	100,00	100,00	100,00
13,69	14,56	13,93	14,24	14,10	14,54	14,82	14,82	14,54	16,09
2,94	2,07	2,70	2,39	2,53	2,09	1,81	1,81	2,09	0,54

IT91	IT92	IT93	IT04	IT94	IT95	IT96	IT97	IT03	IT98
62,37	52,44	56,43	61,85	59,27	55,27	42,08	55,11	52,00	60,62
4,16	4,47	3,92	*3,36*	**2,91**	4,26	*3,61*	7,01	6,22	6,93
0,81	**0,96**	**0,98**	**0,56**	**1,06**	**0,83**	1,76	2,46	2,19	*1,81*
0,35	**0,95**	**1,16**	**0,40**	**0,30**	**0,38**	2,17	1,66	1,53	**1,07**
0,10	**0,06**	**0,04**	**0,03**	**0,04**	**0,08**	**0,14**	**0,12**	**0,11**	**0,09**
13,73	22,73	19,70	15,82	18,73	20,77	32,46	15,25	19,67	*10,88*
16,63	**16,63**	**16,63**	**16,63**	**16,63**	**16,63**	**16,63**	**16,63**	**16,63**	**16,63**
1,76	1,63	**0,95**	*1,24*	**0,94**	1,68	**0,80**	**1,16**	*1,28*	1,62
0,04	**0,07**	**0,14**	**0,09**	**0,08**	**0,06**	**0,22**	*0,43*	**0,29**	**0,25**
0,05	**0,06**	**0,06**	**0,03**	**0,04**	**0,06**	**0,13**	**0,17**	**0,07**	**0,10**
100,00	100,00	100,00	100,00	100,00	100,00	100,00	100,00	100,00	100,00
15,70	15,58	15,63	15,77	15,91	15,57	15,78	15,15	15,25	15,01
0,93	1,05	1,00	0,86	0,72	1,06	0,85	1,48	1,38	1,62

	IT10	IT99	IT100	IT101	IT102	IT103	IT104
SiO_2	60,34	*62,47*	61,77	60,61	44,41	11,87	15,38
Al_2O_3	7,20	7,19	8,20	8,26	8,29	1,31	*1,50*
$Fe_2O_3(t)$	*1,71*	**1,57**	2,35	2,93	3,38	**0,99**	**1,02**
MgO	*1,18*	**1,15**	2,05	2,20	7,74	21,60	20,10
MnO	**0,09**	**0,06**	**0,06**	**0,08**	**0,07**	**0,04**	**0,03**
CaO	*10,81*	**8,99**	**7,01**	**7,39**	17,27	47,10	44,92
Na_2O	**16,63**	**16,63**	**16,63**	**16,63**	**16,63**	**16,63**	**16,63**
K_2O	1,72	1,69	1,60	*1,41*	1,56	**0,27**	**0,29**
TiO_2	**0,23**	**0,17**	**0,25**	**0,38**	*0,55*	**0,13**	**0,09**
P_2O_5	**0,08**	**0,07**	**0,08**	**0,12**	**0,11**	**0,05**	**0,05**
Total	100,00	100,00	100,00	100,00	100,00	100,00	100,00
Added Na_2O	14,93	14,91	14,77	14,95	15,41	16,36	16,31
Sand-derived Na_2O	1,70	1,72	1,86	1,68	1,22	0,27	0,32

Appendix C

Calculated glass compositions after raising the CaO levels of the sands containing insufficient lime to 7.48%, the average CaO content of Roman natron glass (Foster and Jackson, 2009)

Bold: values within compositional ranges (see Table 2.1).
Italic: values within three times the standard deviation. All results are in wt%.

	SP46	SP45	SP44	SP43	SP37	SP36	SP31	SP30
SiO_2	**70,60**	**72,81**	**74,04**	**72,74**	**73,03**	61,46	57,94	51,44
Al_2O_3	**2,91**	*1,48*	1,01	1,79	1,28	7,06	10,27	16,99
$Fe_2O_3(t)$	**1,10**	**0,81**	**0,37**	**0,40**	**0,67**	3,63	4,72	4,00
MgO	**0,29**	**0,13**	**0,05**	**0,12**	**0,64**	2,73	1,18	2,36
MnO	**0,03**	**0,01**	**0,00**	**0,01**	**0,01**	**0,12**	**0,11**	**0,10**
CaO	**7,48**	**7,48**	**7,48**	**7,48**	**7,48**	**7,48**	**7,48**	**7,48**
Na_2O	**16,63**	**16,63**	**16,63**	**16,63**	**16,63**	**16,63**	**16,63**	**16,63**
K_2O	**0,80**	**0,54**	**0,36**	**0,76**	*0,17*	**0,61**	**1,17**	**0,53**
TiO_2	**0,14**	**0,07**	**0,04**	**0,05**	**0,06**	**0,22**	0,42	0,42
P_2O_5	**0,03**	**0,03**	**0,02**	**0,02**	**0,02**	**0,04**	**0,09**	**0,06**
Total	100,00	100,00	100,00	100,00	100,00	100,00	100,00	100,00
Added Na_2O	16,08	16,36	16,47	16,12	16,36	16,15	16,03	16,24
Sand-derived Na_2O	0,55	0,27	0,16	0,51	0,27	0,48	0,60	0,39
Added CaO	5,96	5,47	6,75	5,12	4,99	4,85	2,80	1,92
Sand-derived CaO	1,52	2,01	0,73	2,36	2,49	2,63	4,68	5,56

	FH10	FH09	SP03	FR03	FR04	FR13	FR16
SiO_2	**65,60**	**64,64**	58,97	*62,75*	*64,45*	**72,93**	**70,21**
Al_2O_3	7,03	7,85	9,07	7,91	7,01	*1,47*	**3,29**
$Fe_2O_3(t)$	**0,44**	**1,14**	4,01	*1,73*	**1,48**	**0,13**	**1,28**
MgO	**0,11**	**0,26**	1,41	**0,57**	**0,41**	**0,25**	**0,21**
MnO	**0,01**	**0,01**	**0,04**	**0,02**	**0,02**	**0,00**	**0,01**
CaO	**7,48**	**7,48**	**7,48**	**7,48**	**7,48**	**7,48**	**7,48**
Na_2O	**16,63**	**16,63**	**16,63**	**16,63**	**16,63**	**16,63**	**16,63**
K_2O	2,66	1,90	1,86	2,65	2,35	**1,05**	**0,63**
TiO_2	**0,02**	**0,08**	*0,42*	**0,20**	**0,13**	**0,04**	**0,23**
P_2O_5	**0,03**	**0,03**	**0,11**	**0,06**	**0,04**	*0,01*	**0,03**
Total	100,00	100,00	100,00	100,00	100,00	100,00	100,00
Added Na_2O	14,91	13,88	15,25	15,33	15,48	16,00	15,69
Sand-derived Na_2O	1,72	2,75	1,38	1,30	1,15	0,63	0,94
Added CaO	6,84	5,94	5,06	6,99	6,57	3,75	7,09
Sand-derived CaO	0,64	1,54	2,42	0,49	0,91	3,73	0,39

SP28	SP26	SP22	SP20	FH08	FH03	FH01	FH02	FH12	FH11
67,84	68,12	69,29	71,65	64,80	63,48	65,80	65,40	67,89	68,02
4,18	3,55	3,28	2,18	6,65	9,04	7,09	7,15	5,23	5,18
2,28	2,13	1,64	1,07	1,13	0,63	0,63	0,64	0,44	0,47
0,65	1,11	0,92	0,41	0,66	0,18	0,13	0,14	0,05	0,06
0,03	0,04	0,03	0,01	0,02	0,01	0,02	0,01	0,01	0,01
7,48	7,48	7,48	7,48	7,48	7,48	7,48	7,48	7,48	7,48
16,63	16,63	16,63	16,63	16,63	16,63	16,63	16,63	16,63	16,63
0,62	0,59	0,52	0,40	2,45	2,48	2,18	2,49	2,24	2,11
0,23	0,30	0,17	0,12	0,14	0,04	0,03	0,05	0,02	0,02
0,07	0,05	0,04	0,04	0,04	0,03	0,02	0,02	0,02	0,02
100,00	100,00	100,00	100,00	100,00	100,00	100,00	100,00	100,00	100,00
16,02	16,13	16,12	16,33	15,50	14,42	14,82	14,97	15,46	15,43
0,61	0,50	0,51	0,30	1,13	2,21	1,81	1,66	1,17	1,20
5,34	2,77	2,44	6,30	2,47	5,84	6,75	6,96	7,17	7,16
2,14	4,71	5,04	1,18	5,01	1,64	0,73	0,52	0,31	0,32

FR17	IT14	IT15	IT16a	IT16b	IT17	IT05	IT01	IT47	IT51
67,25	63,37	62,36	44,88	45,11	53,87	60,62	71,63	66,61	57,58
6,08	6,76	6,38	7,49	5,68	6,25	8,26	2,41	5,22	10,56
1,19	2,35	2,64	10,71	6,23	5,12	2,71	0,74	1,61	3,52
0,36	1,51	3,16	11,04	17,70	9,50	1,45	0,18	0,57	1,60
0,03	0,04	0,05	0,13	0,11	0,06	0,04	0,02	0,04	0,05
7,48	7,48	7,48	7,48	7,48	7,48	7,48	7,48	7,48	7,48
16,63	16,63	16,63	16,63	16,63	16,63	16,63	16,63	16,63	16,63
0,61	1,29	0,87	0,79	0,61	0,71	2,43	0,65	1,69	2,07
0,34	0,49	0,39	0,80	0,40	0,33	0,33	0,26	0,12	0,43
0,03	0,09	0,05	0,05	0,04	0,04	0,05	0,01	0,04	0,09
100,00	100,00	100,00	100,00	100,00	100,00	100,00	100,00	100,00	100,00
14,67	14,88	15,05	15,52	15,85	14,93	14,95	16,17	15,49	14,89
1,96	1,75	1,58	1,11	0,78	1,70	1,68	0,46	1,14	1,74
6,50	5,39	4,17	2,84	3,51	5,41	2,79	6,77	4,48	6,00
0,98	2,09	3,31	4,64	3,97	2,07	4,69	0,71	3,00	1,48

	IT52	IT53	IT54	IT55	IT56	IT62	IT63	IT72	IT73	IT74
SiO_2	*64,05*	*63,54*	*63,34*	*64,45*	*63,73*	**74,09**	*74,77*	59,10	56,67	59,6
Al_2O_3	8,19	8,78	8,58	7,78	7,03	0,43	0,33	8,70	10,84	9,27
$Fe_2O_3(t)$	**0,86**	**0,90**	**1,26**	**0,81**	2,39	**0,83**	**0,46**	4,78	4,44	3,79
MgO	**0,30**	**0,30**	**0,42**	**0,29**	0,78	0,28	**0,11**	*1,23*	1,58	**0,77**
MnO	**0,01**	**0,02**	**0,02**	**0,02**	**0,03**	0,02	0,02	**0,06**	**0,09**	**0,27**
CaO	**7,48**	**7,48**	**7,48**	**7,48**	**7,48**	7,48	7,48	**7,48**	**7,48**	7,48
Na_2O	**16,63**	**16,63**	**16,63**	**16,63**	**16,63**	16,63	16,63	**16,63**	**16,63**	16,6
K_2O	2,32	2,19	2,06	2,39	1,65	*0,13*	*0,12*	1,61	1,59	1,76
TiO_2	**0,10**	**0,11**	**0,14**	**0,09**	0,21	**0,07**	**0,06**	**0,33**	0,56	**0,31**
P_2O_5	**0,05**	**0,06**	**0,06**	**0,05**	0,07	**0,03**	**0,02**	**0,09**	0,12	**0,12**
Total	100,00	100,00	100,00	100,00	100,00	100,00	100,00	100,00	100,00	100,0
Added Na_2O	14,81	14,20	14,27	14,49	14,96	16,37	15,97	15,44	14,59	14,99
Sand-derived Na_2O	1,82	2,43	2,36	2,14	1,67	0,26	0,66	1,19	2,04	1,64
Added CaO	5,58	6,00	5,94	4,66	6,01	2,68	1,95	6,46	5,29	5,50
Sand-derived CaO	1,90	1,48	1,54	2,82	1,47	4,80	5,53	1,02	2,19	1,98

	IT75	IT76	IT77	IT78	IT79	IT80	IT81	IT82	IT83
SiO_2	59,65	59,16	60,02	61,38	59,44	*62,58*	*62,34*	61,07	59,76
Al_2O_3	10,79	10,08	10,79	9,82	11,20	9,40	8,72	8,92	10,18
$Fe_2O_3(t)$	2,42	3,12	*1,73*	**1,51**	*1,90*	**0,94**	1,62	2,65	2,71
MgO	**0,86**	**1,09**	**0,58**	**0,48**	**0,80**	0,28	**0,50**	**0,88**	**1,08**
MnO	**0,05**	**0,05**	**0,02**	**0,02**	**0,03**	0,02	**0,03**	**0,03**	**0,05**
CaO	**7,48**	**7,48**	**7,48**	**7,48**	**7,48**	7,48	**7,48**	**7,48**	7,48
Na_2O	**16,63**	**16,63**	**16,63**	**16,63**	**16,63**	16,63	**16,63**	**16,63**	16,63
K_2O	1,73	1,91	2,53	2,45	2,20	2,51	2,43	1,99	1,60
TiO_2	**0,30**	**0,40**	**0,17**	**0,16**	**0,25**	0,09	**0,19**	**0,25**	*0,43*
P_2O_5	**0,09**	**0,09**	**0,05**	**0,06**	**0,07**	0,07	**0,06**	**0,09**	**0,07**
Total	100,00	100,00	100,00	100,00	100,00	100,00	100,00	100,00	100,00
Added Na_2O	13,69	14,56	13,93	14,24	14,10	14,54	14,82	14,82	14,54
Sand-derived Na_2O	2,94	2,07	2,70	2,39	2,53	2,09	1,81	1,81	2,09
Added CaO	6,02	6,39	6,62	6,57	5,60	5,33	5,89	6,35	5,27
Sand-derived CaO	1,46	1,09	0,86	0,91	1,88	2,15	1,59	1,13	2,21

References

Acquafredda, P., Fornelli, A., Piccarreta, G., Summa, V., 1997. Provenance and tectonic implications of heavy minerals in Pliocene-Pleistocene siliciclastic sediments of the southern Apennines, Italy. *Sedimentary Geology 113*, 149-159.

Aerts, A., Janssens, K., Velde, B., Adams, F., Wouters, H., 2000. Analysis of the Composition of Glass Objects from Qumrân, Israel, in: Nenna, M.D. (ed.), *La Route Du Verre. Travaux de la Maison de l'Orient Méditerranéen No. 33*, Lyon, 113–121.

Aerts, A., Velde, B., Janssens, K., Dijkman, W., 2003. Change in silica sources in Roman and post-Roman glass. *Spectrochimica Acta part B 58*, 659-667.

Agenzia per la Protezione dell'Ambiente e per i Servizi Tecnici, 2004. *Carta geologica d'Italia. Scala 1:1,000,000*. S.E.L.C.A. Florence, Italy.

Aloisi, J.C., Duboul-Razavet, C., 1974. Deux exemples de sédimentation deltaïque actuelle en Méditerranée: les deltas du Rhone et de l'Ebre. *Bulletin Centre Recherche Pau, S.N.P.A. 8*, 227-240.

Amorosi, A., Milli, S., 2001. Late Quaternary depositional architecture of Po and Tevere river deltas (Italy) and worldwide comparison with coeval deltaic successions. *Sedimentary Geology 144*, 357-375.

Antonioli, F., Cremona, G., Immordino, F., Puglisi, C., Romagnoli, C., Silenzi, S., Valpreda, E., Verrubbi, V., 2002. New data on the Holocenic sea-level rise in NW Sicily (Central Mediterranean Sea). *Global and Planetary Change 34*, 121-140.

Arletti, R., Dalconi, M.C., Quartieri, S., Triscari, M., Vezzalini, G., 2006. Roman coloured and opaque glass: a chemical and spectroscopic study. *Applied Physics A: Materials Science & Processing 83*, 239-245.

Artignan, D., Nauchbaur, A., 2007. *Mine Pb–Zn–Ag des Bormettes et du Verger Concessions des Bormettes et de La Londe-les-Maures (Var)*. Unpublished Report, Bureau de Recherches Géologiques et Minières (BRGM), RP-55548-FR.

Bamford, C.R., 1977. Colour generation and control in glass. *Glass Science and Technology 2*, Elsevier, Amsterdam.

Banner, J.L., 2004. Radiogenic isotopes: systematics and applications to earth surface processes and chemical stratigraphy. *Earth Science Reviews 65*, 141-194.

Barag, D., 1987. Recent Important Epigraphic Discoveries Related to the History of Glasmaking in the Roman Period, in: *Annales de l'Association Internationale Pour l'Histoire Du Verre* 10 (Madrid and Segovia 1985), 109–116.

Bateson, H.M., Turner, W.E.S., 1939. A note on the solubility of sodium chloride in a soda-lime-silica glass. *Journal of the Society of Glass Technology 23*, 265-267.

Baxter, M.J., Cool, H.E.M., Jackson, C.M., 2005. Further studies in the compositional variability of colourless Romano-British vessel glass. *Archaeometry 47*, 47-68.

Bellotti, P., Caputo, C., Davoli, L., Evangelista, S., Garzanti, E., Pugliese, F., Valeri, P., 2004. Morpho-sedimentary characteristics and Holocene evolution of the emergent part of the Ombrone River delta (southern Tuscany). *Geomorphology 61*, 71-90.

Best, M.G. 2003. *Igneous and metamorphic petrology*. Blackwell Publishing, Malden, MA, 729.

Bingham, P.A., Jackson, C.M., 2008. Roman blue-green bottle glass: chemical-optical analysis and high temperature viscosity modelling. *Journal of Archaeological Science 35*, 302–309.

Bostock, J., Riley, H.T., 1857. *The Natural History of Pliny. Vol.* VI. Henry G. Bohn, York Street, Covent Garden, London. 379-381.

Brems, D., 2012. *Mineralogy and Geochemistry of Mediterranean Sand Deposits as a Raw Material for Roman Natron Glass Production*. PhD Thesis, KU Leuven.

Brems, D., Degryse, P., 2014. Trace element analysis in provenancing Roman glass-making. *Archaeometry DOI: 10.1111/arcm.12063*.

Brems, D., Degryse, P., Ganio, M., Boyen, S., 2012a. The production of Roman glass with western Mediterranean sand raw materials: preliminary results. *Glass Technology: European Journal of Glass Science and Technology A 53*, 129-138.

Brems, D., Boyen, S., Ganio, M., Degryse, P., Walton, M., 2012b. Mediterranean sand deposits as a raw material for glass production in Antiquity. *Annales du 18e Congrès de l'Association Internationale pour l'Histoire du Verre*, Thessaloniki, 120-127.

Brems, D., Degryse, P., Hasendoncks, F., Gimeno, D., Silvestri, A., Vassilieva, E., Luypaers, S., Honings, J., 2012c. Western Mediterranean sand deposits as a raw material for Roman glass production. *Journal of Archaeological Science 39*, 2897-2907.

Brems, D., Ganio, M., Latruwe, K., Balcaen, L., Carremans, M., Gimeno, D., Silvestri, A., Vanhaecke, F., Muchez, P., Degryse, P., 2013a. Isotopes on the beach, Part 1: Strontium isotope ratios as a provenance indicator for lime raw materials used in Roman glassmaking. *Archaeometry 55*, 214-234.

Brems, D., Ganio, M., Latruwe, K., Balcaen, L., Carremans, M., Gimeno, D., Silvestri, A., Vanhaecke, F., Muchez, P., Degryse, P., 2013b. Isotopes on the beach, Part 2: Neodymium isotopic analysis for provenancing Roman glassmaking. *Archaeometry 55*, 449-464.

BRGM (Bureau de Recherches Géologiques et Minières), 2003. Carte géologique de la France à l'échelle du millionième. 6ᵉ edition révisée.

Brill, R.H., 1988. Scientific investigations of the Jalame glass and related finds. In: Weinberg, G.D. (ed.) *Excavations at Jalame. Site of glass factory in Late Roman Palestine*. Missouri Press, Columbia, 257-294.

Brill, R.H., 1999. *Chemical analyses of early glasses*. Corning Museum of glass, Corning, New York.

Brocchini, F., Principe, C., Castratori, D., Laurenzi, M.A., Goria, L., 2001. Quaternary evolution of the southern sector of the Campanian Plain and early Somma-Vesuvius activity: insights from the Trecase well. *Mineralogy and Petrology 73*, 67-91.

Burke, W.H., Denison, R.E., Hetherington, E.A., Koepnick, R.B., Nelson, H.F., Otto, J.B., 1982. Variation of seawater $^{87}Sr/^{86}Sr$ throughout Phanerozoic time. *Geology 10*, 516-519.

Cable, M., Bower, C., 1965. The homogeneity of small-scale laboratory glass melts. *Glass Technology 6*, 197-205.

Caldara, M., Centenaro, E., Mastronuzzi, G., Sansò, P., Sergio, A., 1998. Features and present evolution of Apulian Coast (Southern Italy). *Journal of Coastal Research 26*, 55-64.

Camardo, D., 2007. Archaeology and conservation at Herculaneum: From the Maiuri Campaign to the Herculaneum Conservation Project. *Conservation and Management of Archaeological Sites 8*, 205-214.

Cameron, A., 1993. *The Later Roman Empire, AD 284-430*. Harvard University Press, Cambridge, MA.

Cavazza, W., Zuffa, G.G., Camporesi, C., Ferretti, C., 1993. Sedimentary recycling in a temperate climate drainage basin (Senio River, north-central Italy): Composition of source rock, soil profiles, and fluvial deposits. *Geological Society of America Special Papers 284*, 247-261.

Chopinet, M.-H., Gouillart, E., Papin, S., Toplis, M.J., 2010. Influence of limestone grain size on glass homogeneity. Glass Technology: *European Journal of Glass Science and Technology A 51*, 116-122.

Conticelli, S., D'Antonio, M., Pinarelli, L., Civetta, L., 2002. Source contamination and mantle heterogeneity in the genesis of Italian potassic and ultrapotassic volcanic rocks: Sr-Nd-Pb isotope data from Roman Province and Southern Tuscany. *Mineralogy and Petrology 74*, 189-222.

Cosyns, P., 2001-2002. *Bulletin de l'Association Francaise pur l'Archéologie du Verre, 8*.

Cosyns, P., Martens, M., 2002-2003. *Bulletin de l'Association Francaise pour l'Archéologie du Verre, 34-37*.

Critelli, S., Le Pera, E., Ingersoll, R.V., 1997. The effects of source lithology, transport, deposition and sampling scale on the composition of southern California sand. *Sedimentology 44*, 653-671.

Critelli, S., Arribas, J., Le Pera, E., Tortosa, A., Marsaglia, K.M., Latter, K.K., 2003. The recycled orogenic sand provenance from an uplifted thrust belt, Betic Cordillera, Southern Spain. *Journal of Sedimentary Research 73*, 72-81.

Cullers, R., Chaudhuri, S., Kilbane, N., Koch, R., 1979. Rare-earths in size fractions and sedimentary rocks of Pennsylvanian-Permian age from the mid-continent of the USA. *Geochimica et Cosmochimica Acta 43*, 1285-1301.

Currie, K.J., 2008. *Analytical elemental fingerprinting of natron and its detection in ancient Egyptian mummified remains*. Unpublished PhD Thesis, University of Manchester, UK.

De Francesco, A.M., Scarpelli, R., Barca, D., Ciarallo, A., Buffone, L., 2010. Preliminary chemical characterisation of Roman glass from Pompeii. Period. di Mineral. 79, 11–19.

de Groot, T., 2006. Resultaten van de opgraving van een Romeins tumulusgraf in Bocholtz (Simpelveld) (in Dutch), *Rapportage Archeologische Monumentenzorg 127*, Rijksdienst voor Oudheidkundig Bodemonderzoek, Amersfoort.

De Muynck, D., Huelga-Suarez, G., Van Heghe, L., Degryse, P., Vanhaecke, F., 2009. Systematic evaluation of a strontium-specific extraction chromatographic resin for obtaining a purified Sr fraction with quantitative recovery from complex and Ca-rich matrices. *Journal of Analytical Atomic Spectrometry 24*, 1498-1510.

Degeest, R., 2000. *The Common Wares of Sagalassos. Typology and chronology. Studies in Eastern Mediterranean Archaeology 3*. Brepols, Turnhout.

Degryse, P., Schneider, J., 2008. Pliny the Elder and Sr-Nd isotopes: tracing the provenance of raw materials for Roman glass production. *Journal of Archaeological Science 35*, 1993-2000.

Degryse, P., Shortland, A., 2009. Trace elements in provenancing raw materials for Roman glass production. *Geologica Belgica 12*, 135-143.

Degryse, P., Schneider, J., Poblome, J., Muchez, Ph., Haack, U., Waelkens, M., 2005. Geochemical study of Roman to Byzantine Glass from Sagalassos, Southwest Turkey. *Journal of Archaeological Science 32*, 287-299.

Degryse, P., Schneider, J., Haack, U., Lauwers, V., Poblome, J., Waelkens, M., Muchez, Ph., 2006a. Evidence for glass 'recycling' using Pb and Sr isotopic ratios and Sr-mixing lines: the case of early Byzantine Sagalassos. *Journal of Archaeological Science 33*, 494-501.

Degryse, P., Schneider, J., Lauwers, V., De Muynck, D., Vanhaecke, F., Waelkens, M., Muchez, Ph., 2006b. Sr and Nd isotopic evidence in the primary provenance determination of Roman glass from Sagalassos (SW Turkey). *Annales du 17e Congrès de l'Association Internationale pour l'Histoire du Verre*, Anvers, 564-570.

Degryse, P., Schneider, J., Lauwers, V., De Muynck, D., Vanhaecke, F., Waelkens, M., Muchez, Ph., 2006c. Sr and Nd isotopic evidence in the primary provenance determination of Roman glass from Sagalassos (SW Turkey). *Annales du 17e Congrès de l'Association Internationale pour l'Histoire du Verre*, Anvers, 564-570.

Degryse, P., Schneider, J., Lauwers, V., Brems, D., 2008. Sr-Nd isotopic analysis of glass from Sagalassos (SW Turkey). *Journal of Cultural Heritage 9*, e47-e49.

Degryse, P., Henderson, J., Hodgins, G., 2009a. Isotopes in vitreous materials, a state-of-the-art and perspectives. In: Degryse, P., Henderson, J., Hodgins, G. (eds.) Isotopes in Vitreous Materials. *Studies in Archaeological Sciences*. Leuven University Press, 15-30.

Degryse, P., Schneider, J., Lauwers, V., Henderson, J., Van Daele, B., Martens, M., Huisman, H.D.J., De Muynck, D., Muchez, Ph., 2009b. Neodymium and strontium isotopes in the provenance determination of primary natron glass production. In: Degryse, P., Henderson, J., Hodgins, G. (eds.) Isotopes in Vitreous Materials. *Studies in Archaeological Sciences*. Leuven University Press, 53-72.

Degryse, P., Shortland, A., De Muynck, D., Vandeweghe, L., Scott, R., Vanhaecke, F., 2010. Limitations to the provenance determination of plant ash glasses using strontium isotopes. *Journal of Archaeological Science 37*, 3129–3135.

Degryse, P., Scott, R.B., Brems, D., 2014. The archaeometry of ancient glassmaking: reconstructing ancient technology and the trade of raw materials. *Perspective, Revue de l'Institut national d'histoire de l'art (INHA)*, 2014-2.

DePaolo, D.J., 1988. *Neodymium isotope geochemistry: an introduction*. Springer, Berlin, 187.

Devulder, V., Degryse, P., Vanhaecke, F., 2013a. Development of a Novel Method for Unraveling the Origin of Natron Flux Used in Roman Glass Production Based on B Isotopic Analysis via Multicollector Inductively Coupled Plasma Mass Spectrometry. *Analytical Chemistry 85* (24), 12077-12084.

Devulder, V., Lobo, L., Van Hoecke, K., Degryse, P., Vanhaecke, F., 2013b. Common analyte internal standardization as a tool for correction formass discrimination in multi-collector inductively coupled plasma-mass spectrometry. *Spectrochimica Acta B, Atomic Spectroscopy 89*, 20-29.

Devulder, V., Vanhaecke, F., Shortland A., Mattingly, D., Jackson, C., Degryse, P., 2014. Boron isotopic composition as a provenance indicator for the flux raw material in Roman natron glass. *Journal of Archaeological Science 46*, 107-113

Diels, M., Brems, D., Degryse, P., 2014. *Suitability of modern beach sands as a raw material for Roman glass production: case study in SE Italy (Apulia and Basilicata)*. Unpublished Master Thesis, KU Leuven.

Dijkman, W., 1993, La terre sigillée décorée à la molette à motifs chrétiens dans la stratigraphie maastrichtoise (Pays-Bas) et dans le nord-ouest de l'Europe, in: *Gallia 49*, 129-172.

Dotsika, E., Poutoukis, D., Tzavidopoulos, I., Maniatis, Y., Ignatiadou, D., Raco, B., 2009. A natron source at Pikrolimni Lake in Greece? Geochemical evidence. *Journal of Geochemical Exploration, 103(2-3)*,133–143.

Duplaix, S., 1972. Les minéraux lourds de sables de plages et de canyons sous-marins de la Méditerranée Française. In: Stanley, D.J., Kelling, G., Weiler, Y. (eds.) *The Mediterranean Sea: A natural sedimentation laboratory*. Dowden, Hutchinson & Ross, Inc. Stroudsburg, Pennsylvania. 293-303.

Eremin, K., Degryse, P., Erb-Satullo, N., Ganio, M., Greene, J., Shortland, A., Walton, M., Stager, L., 2011. Amulets and Infant Burials: Glass beads from the Carthage Tophet. *Proceedings of the SEM and microanalysis in the study of historical technology, materials and conservation Conference*, London, UK, 9-10 September 2010.

Erim, K.T., Reynolds, J., 1973. The Aphrodisias Copy of Diocletian's Edict on Maximum Prices. *J. Rom. Stud. 63*, 99–110.

Fabri, M., Degryse, P., 2013. *Chemische merkers voor natron zouten (Chemical markers for natron salts)* (in Dutch) Unpublished Bachelor Thesis, KU Leuven.

Faure, G., Mensing, T.M., 2005. *Isotopes: principles and applications. 3rd Edition*. John Wiley and Sons, New York, 897.

Féraud, J., 1983. Ore veins linked to old emersion surfaces in the crystalline basement of Provence and of the external belt of the French–Italian Alps. In: Schneider, H.J. (ed.) *Mineral deposits of the Alps and of the Alpine Epoch in Europe. Society for Geology Applied to Mineral Deposits Special Publication 3*. Springer, Prague, 128-135.

Fleming, S.J., Swann, C.P., 1999. Roman mosaic glass: a study of production processes, using PIXE spectrometry. *Nucl. Instruments Methods Phys. Res. B 150*, 622–627.

Flemming, N.C., 1969. Archaeological evidence for eustatic change of sea level and earth movements in the western Mediterranean during the last 2000 years. *Geological Society of America Special Paper 109*, 125.

Foster, G.L., Vance, D., 2006. In situ Nd isotopic analysis of geological materials by laser ablation MC-ICPMS. *Journal of Analytical Atomic Spectrometry 21*, 288-296.

Foster, H.E., Jackson, C.M., 2009. The composition of 'naturally coloured' late Roman vessel glass from Britain and the implications for models of glass production and supply. *Journal of Archaeological Science 36*, 189-204.

Foster, H.E., Jackson, C.M., 2010. The composition of late Romano-British colourless vessel glass: glass production and consumption. *Journal of Archaeological Science 37*, 3068-3080.

Foy, D., Sennequier, G., 1989. *À travers le verre du moyen âge à la renaissance. Musées et Monuments departementaaux de la Seine-Maritime*, Rouen.

Foy, D., Nenna, M.-D., 2001. Tout feu tout sable. *Mille ans de verre antique dans le Midi de la France*. Édisud, Marseille.

Foy, D., Fontaine, S.D., 2007. L'épave Ouest-Embiez 1, Var, le commerce maritime du verre brut et manufacturé en Méditerranée occidentale dans l'Antiquité. *Revue archéologique de Narbonnaise*, 235-268.

Foy, D., Vichy, M., Picon, M., 2000. Lingots de verre en Méditerranée occidentale. *Annales du 14e Congrès de l'Association Internationale pour l'Histoire du Verre*, Amsterdam, 51-57.

Foy, D., Picon, M., Vichy, M., Thirion-Merle, V., 2003. Caractérisation des verres de la fin de l'Antiquité en Méditerranée occidentale: l'émergence de nouveaux courants commerciaux. In: Foy, D., Nenna, M.D. (eds.) *Échanges et Commerce du Verre dans le Monde Antique. Monographies Instumentum 24*. Éditions Monique Mergoil, Montagnac, 41-85.

Freestone, I.C., Degryse, P., Shepherd, J., Gorin-Rosen, Y., Schneider, J., n.d. Neodymium and Strontium Isotopes Indicate a Near Eastern Origin for Late Roman Glass in London. *Journal of Archaeological Science*.

Freestone, I.C., 2001. Primary Glass Sources in the Mid First Millennium AD, in: *Annales Du 15e Congrès de l'Association Internationale Pour l'Histoire Du Verre (AIHV)*. USA, 111–115.

Freestone, I.C., 2005. The Provenance of Ancient Glass through Compositional Analysis. *Mater. Res. Soc. Symp. Proc. 852*, 008.1.1–008.1.14.

Freestone, I.C., 2006. Glass production in Late Antiquity and the Early Islamic period: a geochemical perspective. In: Maggetti, M., Messiga, B. (eds.) *Geomaterials in Cultural Heritage*. Geological Society of London, Special Publications 257, 201-216.

Freestone, I.C., 2008. Pliny on Roman Glassmaking. In: Matinón-Torres, M., Rehren, T. (eds.) *Archaeology, History and Science. Integrating Approaches to Ancient Materials* (UCL Institute of Archaeology Publications), Oxford, 77-100.

Freestone, I.C., Gorin-Rosen, Y., Hughes, M.J., 2000. Primary glass from Israel and the production of glass in the late antiquity and the early Islamic period. In: Nenna, M.D. (ed.) La route du verre. Ateliers primaires et secondaires de verriers du second millénaire avant J.C. au Moyen Age. *Travaux de la Maison de l'Orient Méditerranéen 33*, Lyon, 65-83.

Freestone, I.C., Ponting, M., Hughes, J., 2002a. Origins of Byzantine glass from Maroni Petrera, Cyprus. *Archaeometry 44*, 257-272.

Freestone, I.C., Greenwood, R., Gorin-Rosen, Y., 2002b. Byzantine and early Islamic glassmaking in the Eastern Mediterranean: production and distribution of primary glass. In: Kordas, G. (ed.) *Proceedings of the first international conference on Hyalos-Vitrum-Glass. History, technology and conservation of glass and vitreous materials in the Hellenic World*, 1-4 April 2001, Rhodes, Greece. Glasnet Publications, Athens, 167-174.

Freestone, I.C., Leslie, K.A., Thirlwall, M., Gorin-Rosen, Y., 2003. Strontium isotopes in the investigation of early glass production: Byzantine and early Islamic glass from the Near East. *Archaeometry 45*, 19-32.

Freestone, I.C., Wolf, S., Thirlwall, M., 2005. The production of HIMT glass: elemental and isotopic evidence. *Annales du 16e Congrès de l'Association Internationale pour l'Histoire de Verre*, London, 153-157.

Freestone, I.C., Jackson-Tal, R.E., Tal, O., 2008a. Raw Glass and the Production of Glass Vessels at Late Byzantine Apollonia-Arsuf, Israel. *J. Glass Stud. 50*, 67–80.

Freestone, I.C., Hughes, M.J., Stapleton, C.P., 2008b. The composition and production of Anglo-Saxon glass. In: Evison, V.I. (ed.), *Catalogue of Anglo-Saxon Glass in the British Museum. The British Museum*, London, 29-46.

Freydier, R., Michard, A., De Lange, G., Thomson, J., 2001. Nd isotopic compositions of Eastern Mediterranean sediments: tracers of Nile influence during sapropel S1 formation? *Marine Geology 177*, 45-62.

Frost, C.D., O'Nions, R.K., Goldstein, S.L., 1986. Mass balance for Nd in the Mediterranean Sea. *Chemical Geology 55*, 45-50.

Gandolfi, G., Mordenti, A., Paganelli, L., 1982. Composition and longshore dispersal of sands from the Po and Adige rivers since the pre-Etruscan age. *Journal of Sedimentary Petrology 52*, 797-805.

Ganio, M., 2013. *A 'true'Roman glass: evidence for primary production in Italy*. PhD Thesis, KU Leuven.

Ganio, M., Latruwe, K., Brems, D., Degryse, P., Muchez, P., Vanhaecke, F., 2012a. Sr-Nd isolation procedure for subsequent isotopic analysis using multi-collector ICP – mass spectrometry in the context of provenance studies on archaeological glass. *Journal of Analytical Atomic Spectrometry. DOI: 10.1039/c2ja30154g*

Ganio, M., Boyen, S., Brems, D., Scott, R., Foy, D., Latruwe, K., Molin, G., Silvestri, A., Vanhaecke, F., Degryse, P., 2012b. Trade routes across the Mediterranean: a Sr/Nd isotopic investigation on Roman colourless glass. Glass Technology: *European Journal of Glass Science and Technology A 53*, 217-224.

Ganio, M., Boyen, S., Fenn, T., Scott, R., Vanhoutte, S., Gimeno, D., Degryse, P., 2012c. Roman glass across the Empire: an elemental and isotopic characterization. *Journal of Analytical Atomic Spectrometry 27*, 743-753.

Garzanti, E., Scutellà, M., Vidimari, C., 1998. Provenance from ophiolites and oceanic allochthons: modern beach and river sands from Liguria and the Northern Apennines (Italy). *Ofioliti 23*, 65-82.

Garzanti, E., Canclini, S., Moretti Foggia, F., Petrella, N., 2002. Unravelling magmatic and orogenic provenances in modern sands: the back-arc side of the Apennine thrust-belt (Italy). *Journal of Sedimentary Research 72*, 2-17.

Garzanti, E., Vezzoli, G., Lombardo, B., Andò, S., Mauri, E., Monguzzi, S., Russo, M., 2004. Collision-orogen provenance (Western Alps): detrital signatures and unroofing trends. *Journal of Geology 112*, 145-164.

Gerth, K., Wedepohl, K., Heide, K., 1998. Experimental melts to explore the technique of medieval woodash glass production and the chlorine content of medieval glass types. *Chemie der Erde 58*, 219-232.

Giannetti, B., 2001. Origin of the calderas and evolution of Roccamonfina volcano (Roman Region, Italy). *Journal of Volcanology and Geothermal Research 106*, 301-319.

Gibbins, D.J.L., 1991. The Roman wreck of c. AD 200 at Plemmirio, near Siracusa (Sicily): *third interim report. Int. J. Naut. Archaeol. 20*, 227–246.

Goldstein, S.L., O'Nions, R.K., Hamilton, P.J., 1984. A Sm-Nd isotopic study of atmospheric dusts and particulates from major river systems. *Earth and Planetary Science Letters 70*, 221-236.

Gorin-Rosen, Y., 1995. Hadera, Bet Eli'ezer. *Excavations and Surveys in Israel 13*, 42-43.

Gorin-Rosen, Y., 2000. The ancient glass industry in Israel: summary of finds and new discoveries. In: Nenna, M.D. (ed.) *La route du verre. Ateliers primaires et secondaires du second millénaire avant J.C. au Moyen Age. Travaux de la Maison de l'Orient Méditerranéen 33*, Lyon, 49-64.

Gorin-Rosen, Y., 2012. Remains of a Glass Industry and Glass Finds from Horbat Biz'a. *Atiqot 70*, 49–62.

Gratuze, B., Barrandon, J.-N., 1990. Islamic glass weights and stamps: analysis using nuclear techniques. *Archaeometry 32*, 155–162.

Green, L.R., Hart, F.A., 1987. Colour and chemical composition in ancient glass: An examination of some Roman and Wealden glass by means of ultraviolet-visible-infra-red spectrometry and electron microprobe analysis. *Journal of Archaeological Science 14*, 271-282.

Gromet, L.P., Silver, L.T., 1983. Rare earth element distributions among minerals in a granodiorite and their petrogenetic implications. *Geochimica et Cosmochimica Acta 47*, 925-939.

Grousset, F.E., Biscaye, P.E., 2005. Tracing dust sources and transport patterns using Sr, Nd and Pb isotopes. *Chemical Geology 222*, 149-167.

Grousset, F.E., Biscaye, P.E., Zindler, A., Prospero, J., Chester, R., 1988. Neodymium isotopes as tracers in marine sediments and aerosols: North Atlantic. *Earth and Planetary Science Letters 87*, 367-378.

Grousset, F.E., Parra, M., Bory, A., Martinez, P., Bertrand, P., Shimmield, G., Ellam, R.M., 1998. Saharan wind regimes traced by the Sr-Nd isotopic composition of the Subtropical Atlantic sediments: Last Glacial Maximum vs. today. *Quaternary Science Reviews 17*, 395-409.

Guillén, P., Palanques, A., 1997. A historical perspective of the morphological evolution in the lower Ebro River. *Environmental Geology 30*, 174-180.

Gulson, B.L., 1969. Electron microprobe determination of Zr/Hf ratios in zircons from the Yeoval Diorite Complex, N.S.W., Australia. *Lithos 3*, 17-23.

Hamad, N., Millot, C., Taupier-Letage, I., 2006. The surface circulation in the eastern basin of the Mediterranean Sea. *Scientia Marina 70*, 457-503.

Harding, I.C., Trippier, S., Steele, J., 2004. The provenancing of flint artefacts using palynological techniques, in: Walker E.A., Wenban-Smith F., F., H. (eds.), *Lithics in Action: Papers from the Conference Lithic Studies in the Year 2000*. Oxford: Oxbow Books, 78–88.

Hares, G.B., 1984. Composition and Constitution, in: McLellan, G.W., Shand, E.B. (eds.), *Glass Engineering Handbook*. McGraw-Hill Book Company, New York, 1.3–1.11.

Healy, J.F., 1999. Glass technology. In: Healy, J.F., *Pliny the Elder on Science and Technology*. Oxford University Press, Oxford. 352-358.

Henderson, J., 1985. The raw materials of early glass production. Oxford *Journal of Archaeology 4*, 267-291.

Henderson, J., Evans, J.A., Sloane, H.J., Leng, M.J., Doherty, C., 2005. The use of oxygen, strontium and lead isotopes to provenance ancient glasses in the Middle East. *Journal of Archaeological Science 32*, 665-673.

Henry, F., Jeandel, C., Dupré, B., Minster, J.-F., 1994. Particulate and dissolved Nd in the Western Mediterranean Sea: sources, fates and budget. *Marine Chemistry 45*, 283-305.

Hodges, H., 1981. *Artefacts: An introduction to early materials and technology.* John Baker, London.

Horne J.F., 1895. *The buried cities of Vesuvius: Herculaneum and Pompeii.* Hazell, Watson and Viney, London.

Ignatiadou, D., Dotsika, E., Kouras, A., Maniatis, Y., 2005. Nitrum chalestricum: the natron of Macedonia. In M.D. Nenna, editor, *Annales du 16e Congres de l'Association Internationale pour l'Histoire du Verre*, Nothingham.

Institut Geològic de Catalunya, 2006. *Mapa de grups litològics de Catalunya, 1:250,000.*

Institut Geològic de Catalunya, 2011. *Mapa geològic de Catalunya, 1:300,000. Edició 2.*

Instituto Technológico Geominero de España, 1994. *Mapa Geológico de la Península Ibérica, Baleares y Canarias a escala 1:1.000.000.*

Isings C., 1957. Roman glass from dated finds, *Archaeologica Traiectina II.*

Ixer, R.A., 2007. Foundered or founded on rock- a future for Welsh Provenance Studies. Horwitz, E.P., Dietz, M.L., Fischer, D.E., 1991. SREX: a new process for the extraction and recovery of strontium from acidic nuclear waste streams. *Solvent Extraction and Ion Exchange 9*, 1-25.

Jackson, C.M., 2005. Making colourless glass in the Roman period. *Archaeometry 47*, 763-780.

Jackson, C.M., 2012. On the Provenance of Roman Glasses, in: Liritzis, I., Stevenson, C.M. (eds.), *Obsidian and Ancient Manufactured Glasses*. University of New Mexico Press, Albuquerque, 157–165.

Jackson, C.M., Smedley, J.W., 2004. Medieval and post-medieval glass technology: melting characteristics of some glasses melted from vegetable ash and sand mixtures. *Glass Technology 45*, 36-42.

Jackson, C.M., Baxter, M.J., Cool, H.E.M., 2003a. Identifying Group and Meaning: An investigation of Roman Colourless Glass, in: Foy, D., Nenna, M.D. (eds.), *Échanges et Commerce Du Verre Dans Le Monde Antique. Monographies Instrumentum 24.* Editions Monique Mergoil, Montagnac, 33–39.

Jackson, C.M., Joyner, L., Booth, C.A., Day, P.M., Wager, E.C., Kilikoglou, V., 2003b. Roman glass-making at Coppergate, York? Analytical evidence for the nature of production. *Archaeometry 45*, 435-456.

Jeandel, C., Arsouze, T., Lacan, F., Téchiné, P., Dutay, J.-C., 2007. Isotopic Nd compositions and concentrations of the lithogenic inputs into the ocean: A compilation, with an emphasis on the margins. *Chemical Geology 239*, 156-164.

Johnsson, M.J., 1993. The system controlling the composition of clastic sediments. In: Johnson, M.J., Basu, A. (eds.) Processes controlling the composition of clastic sediments. *Geological Society of America Special Paper 284*, 1-19.

Joukowsky, M. S., 2007. *Petra Great Temple: Archaeological Context of the Remains and Excavations, Vol. 2*, Providence.

Kakhidze, E., 2008. *Meetings of Cultures in the Black Sea region, 8*, 303-332.

Katz, A., Sass, E., Starinsky, A., Holland, H.D., 1972. Strontium behavior in the aragonite-calcite transformation: An experimental study at 40-98°C. *Geochimica et Cosmochimica Acta 36*, 481-496.

Kinsman, D.J.J., 1969. Interpretation of Sr^{2+} concentrations in carbonate minerals and rocks. *Journal of Sedimentary Research 39*, 486-508.

Köpsel, D., 2001. Solubility and Vaporization of Halides. *Proceedings of International Congress on Glass*, Edinburgh, Scotland, 330.

Krom, M.D., Cliff, R.A., Eijsink, L.M., Herut, B., Chester, R., 1999a. The characterisation of Saharan dusts and Nile particulate matter in surface sediments from the Levantine basin using Sr isotopes. *Marine Geology 155*, 319-330.

Krom, M.D., Michard, A., Cliff, R.A., Strohle, K., 1999b. Sources of sediment to the Ionian Sea and western Levantine basin of the Eastern Mediterranean during S-1 sapropel times. *Marine Geology 160*, 45-61.

Lambeck, K., Anzidei, M., Antonioli, F., Benini, A., Esposito, A., 2004. Sea level in Roman time in the Central Mediterranean and implications for recent change. *Earth and Planetary Science Letters 224*, 563-575.

Lambeck, K., Bard, E., 2000. Sea-level change along the French Mediterranean coast for the past 30 000 years. *Earth and Planetary Science Letters 175*, 203-222.

Larue, J.-P., 2008. Effects of tectonics and lithology on long profiles of 16 rivers of the southern Central Massif border between the Aude and the Orb (France). *Geomorphology 93*, 343-367.

Lauwers, V., 2008. *The glass of Sagalassos. A geochemical and typo-chronological interpretation*. PhD Thesis, KU Leuven.

Le Pera, E., Critelli, S., 1997. Sourceland controls on the composition of beach and fluvial sand of the northern Tyrrhenian coast of Calabria, Italy: implications for actualistic petrofacies. *Sedimentary Geology 110*, 81-97.

Lewis, N., Reinhold, M., 1966. *Roman Civilization*. Harper Torchbooks, New York.

Lidiak, E.G., Jolly, W.T., 1996. Rare earth elements in the geological sciences. In: Evans, C.H. (ed.) *Episodes from the history of the rare earth elements*. Kluwer Academic Publishers, Dordrecht. 149-187.

Linn, A.M., DePaolo, D.J., 1993. Provenance controle on the Nd-Sr-O isotopic composition of sandstones: Examples from Late Mesozoic Great Valley forearc basin, California. In: Johnson, M.J., Basu, A. (eds.) Processes controlling the composition of clastic sediments. *Geological Society of America Special Paper 284*, 121-133.

Liquete, C., Arnau, P., Canals, M., Colas, S., 2005. Mediterranean river systems of Andalusia, southern Spain, and associated deltas: a source to sink approach. *Marine Geology 222-223*, 471-495.

Lucas, A., 1932. The Occurrence of Natron in Ancient Egypt. *J. Egypt. Archaeol. 18*, 62–66.

Lucas, A., Harris, J.R., 1962. *Ancient Egyptian Materials and Industries*. Edward Arnold (publishers) LTD, London, fourth edition.

Macklin, M.G., Lewin, J., Woodward, J.C., 1995. Quaternary fluvial systems in the Mediterranean basin. In: Lewin, J., Macklin, M.G., Woodward, J.C. (eds.) *Mediterranean Quaternary River Environments*. Balkema, Rotterdam, 1-25.

Maldonado, A., 1972. El delta de l'Ebro. Estudio sedimentològico y estratigrafico. *Boletìn de Estratigrafìa 1*. Universidad de Barcelona, 474.

Maldonado, A., 1975. Sedimentation, stratigraphy, and development of the Ebro Delta, Spain. In: Broussard, M.L. (ed.) *Delta Models for Exploration*. Houston Geological Society, Houston, Texas, 311-338.

Malone, S., 2010. Assessment of the archaeological remains and an updated project design for excavations at Grandcourt quarry, East Winch, Norfolk. *APS Reports 17/10*.

Marii, F., 2001. *In The Petra Church*; ed. Bikai P., Jordan, Amman: American Center of Oriental Research, 377-383.

Mastronuzzi, G., Sansò, P., 2002. Holocene coastal dune development and environmental changes in Apulia (southern Italy). *Sedimentary Geology 150*, 139-152.

Mastronuzzi, G., Quinif, Y., Sansò, P., Selleri, G., 2007. Middle-Late Pleistocene polycyclic evolution of a stable coastal area (southern Apulia, Italy). *Geomorphology 86,* 393-408.

Matson, F.R., 1951. The Compostion and Working Properties of Ancient Glass. *J. Chem. Educ. 28,* 82–87.

McFarlane, C.R.M., McCulloch, M.M., 2007. Coupling of in-situ Sm-Nd systematics and U-Pb dating of monazite and allanite with applications to crustal evolution studies. *Chemical Geology 245,* 45-60.

McKay, G.A., 1989. Partitioning of rare earth elements between major silicate minerals and basaltic melts. In: Lipin, B.R., McKay, G.A. (eds.) Geochemistry and Mineralogy of Rare Earth Elements. Mineralogical Society of America, *Reviews in Mineralogy 21,* 45-77.

McLennan, S.M., 1989. Rare earth elements in sedimentary rocks: influence of provenance and sedimentary processes. In: Lipin, B.R., McKay, G.A. (eds.) Geochemistry and Mineralogy of Rare Earth Elements. Mineralogical Society of America, *Reviews in Mineralogy 21,* 169-200.

Mirti, P., Lepora, A., Saguì, L., 2000. Scientific analysis of seventh-century glass fragments from the Crypta Balbi in Rome. *Archaeometry 42,* 359-374.

Mirti, P., Davit, P., Gulmini, M., Sagui, L., 2001. Glass fragments from the Crypta Balbi in Rome: The composition of eighth century fragments. *Archaeometry 43,* 491-502.

Morey, G.W., 1938. *The Properties of Glass*. Reinhold Publishing Company, New York.

Morey, G.W., Bowen, N.L., 1925. The Melting Relations of the Soda-Lime-Silica Glasses. *J. Soc. Glas. Technol. 9,* 226–264.

Nardi, L.V.S., Formoso, M.L.L., Jarvis, K., Oliveira, L., Bastos Neto, A.C., Fontana, E., 2012. REE, Y, Nb, U, and Th contents and tetrad effect in zircon from a magmatic-hydrothermal F-rich system of Sn-rare metalecryolite mineralized granites from the Pitinga Mine, Amazonia, Brazil. *Journal of South American Earth Sciences 33,* 34-42.

Nelson, B.W., 1970. Hydrography, sediment dispersal, and recent historical development of the Po river delta, Italy. In: Morgan, J.P., Shaver, R.H. (eds.) Deltaic Sedimentation: Modern and Ancient. *Society of Economic Paleontologists and Mineralogists, Special Publication 15,* 152-184.

Nenna, M.D., 2003. Les ateliers égyptiens à l'époque gréco-romaine, in: Foy, D. (ed.), *Coeur de Verre. Production et Diffusion Du Verre Antique*. Gollion, 32–33.

Nenna, M.D., 2007. Production et commerce du verre à l'époque impériale: nouvelles découvertes et problématiques. *Facta 1,* 125-147.

Nenna, M.D., 2014. Egyptian glass abroad: HIMT glass and its markets. In: Keller, D., Price, J., Jackson, C. (eds.) *Neighbours and successors of Rome. Traditions of Glass Production and Use in Europe and the Middle East in the later First Millenium AD*. Oxbow Books, 177-193.

Nenna, M.D., Vichy, M., Picon, M., 1997. L'Atelier de verrier de Lyon, du Ier siècle après J.-C., et l'origine des verres 'Romains'. *Revue d'Archéométrie 21*, 81-87.

Nenna, M.D., Picon, M., Vichy, M., 2000. Ateliers primaires et secondaires en Egypte à l'époque Gréco-Romaine. In: Nenna, M.D. (ed.) *La route du verre. Ateliers primaires et secondaires du second millénaire avant J.C. au Moyen Age. Travaux de la Maison de l'Orient Méditerranéen 33*, Lyon, 97-112.

Nenna, M.D., Picon, M., Thirion-Merle, V., Vichy, M., 2005. Ateliers primaires du Wadi Natrun: nouvelles découvertes. *Annales du 16e Congrès de l'Association Internationale pour l'Histoire du Verre*, London, 59-63.

O'Hea, M., 2001. *In The Petra Church*; ed. Bikai P., Jordan, Amman: American Center of Oriental Research, 370-376.

Oomkens, E., 1970. Depositional sequences and sand distribution in the post-glacial Rhône delta complex. In: Morgan, J.P., Shaver, R.H. (eds.) *Deltaic Sedimentation: Modern and Ancient. Society of Economic Paleontologists and Mineralogists, Special Publication 15*, 198-212.

Panhuysen, T., Dijkman, W., 1987. Romeinse vondsten van de Scharnweg, in: *Publications de la Société Archéologique et Historique dans le Limbourg, CXXIII*, 212-216.

Pauwels, J., Brems, D., 2014. *Laterale variatie in mineralogische en geochemische karakteristieken van strandzand en de implicaties voor de herkomstbepaling van Romeins glas. (Lateral variation in mineralogical and geochemical properties of beach sands and the implication for provenancing Roman glass)* (in Dutch) Unpublished Bachelor Thesis, KU Leuven.

Paynter, S., 2006. Analyses of colourless Roman glass from Binchester, County Durham. *Journal of Archaeological Science 33*, 1037-1057.

Paynter, S., Dungworth, D., 2011. Recognising frit: experiments reproducing post-Medieval plant ash glass. In: *Proceedings of the 37th International Symposium on Archaeometry*, 13-16 may 2008, Siena, Italy. 133-138.

Perkins, A., 1951. Archaeological News. *Am. J. Archaeol. 55*, 81–100.

Picard, M.D., McBride, E.F., 2007. Comparison of river and beach sand composition with source rocks, Dolomite Alps drainage basin, northeastern Italy. *Geological Society of America Special Papers 420*, 1-12.

Picon, M., Vichy, M., 2003. D'Orient en Occident: l'origine du verre à l'époque romaine et Durant le haut Moyen Âge. In: Foy, D., Nenna, M.D. (eds.) *Echanges et commerce du verre dans le monde antique. Monographies Instumentum 24*. Éditions Monique Mergoil, Montagnac, 17-31.

Pillay, A.E., 2001. Analysis of archaeological artefacts: PIXE, XRF or ICP-MS? *J. Radioanal. Nucl. Chem. 247*, 593–595.

Pinardi, N., Masetti, E., 2000. Variability of the large general circulation of the Mediterranean Sea from observations and modelling: a review. *Palaeogeography, Palaeoclimatology, Palaeoecology 158*, 153-173.

Poblome, J., 1999. *Sagalassos Red Slip Ware. Typology and chronology. Studies in Eastern Mediterranean Archaeology 2*. Brepols, Turnhout.

Pollard, A.M., Heron, C., 2008. *Archaeological Chemistry, Second Edition.* The Royal Society of Chemistry, Cambridge, London.

Poulos, S.E., Collins, M.B., 2002. Fluviatile sediment fluxes to the Mediterranean Sea: a quantitative approach and the influence of dams. In: Jones, S.J., Frostick, L.E. (eds.) Sediment Flux to Basins: Causes, Controls and Consequences. *Geological Society of London, Special Publications 191,* 227-245.

Price, J., 2005. Glass-working and glassworkers in cities and towns, in: MacMahon, A., Price, J. (eds.), *Roman Working Lives and Urban Living.* Oxbow Books, Oxford, 167–190.

Reade, W., 2009. *A compositional study of Late Bronze Age and Iron Age glasses from the Near East.* Unpublished PhD Thesis, The University of Sydney, Austalia.

Rehren, Th., 2000. Rationales in Old World base glass compositions. *Journal of Archaeological Science 27,* 1225-1234.

Rey, T., Lefevre, D., Vella, C., 2009. Deltaic plain development and environmental changes in the Petite Camargue, Rhone Delta, France, in the past 2000 years. *Quaternary Research 71,* 284-294.

Rizzini, A., 1974. Holocene sedimentary cycle and heavy-mineral distribution, Romanga-Marche coastal plain, Italy. *Sedimentary Geology 11,* 17-37.

Robinson, C.E., 1978. *A History of Rome from 753 B.C. to A.D. 410.* Methuen Educational Ltd, London.

Rosenow D. and Rehren Th., 2014. Herding cats – Roman to late Antique glass groups from Bubastis, northern Egypt. *Journal of Archaeological Science, 49,* 170-184.

Rottländer, R.C.A., 1986. The Pliny translation Group of Germany. In: French, R., Greenaway, F. (eds.) *Science in the Early Roman Empire: Pliny the Elder, his sources and influence.* Croom Helm, London, 11-19.

Saguí, L., 2007. Glass in Late Antiquity: The Continuity of Technology and Sources of Supply, in: Lavan, L., Zanini, E., Saranris, A. (eds.), *Technology in Transition A.D. 300-650.* Brill, Boston, 211–231.

Salviulo, G., Silvestri, A., Molin, G., Bertoncello, R., 2004. An archaeometric study of the bulk and surface weathering characteristics of Early Medieval (5th–7th century) glass from the Po valley, northern Italy. *Journal of Archaeological Science 31,* 295-306.

Santisteban, J.I., Schulte, L., 2007. Fluvial networks of the Iberian Peninsula: a chronological framework. *Quaternary Science Reviews 26,* 2738-2757.

Savage, S., Keller, D., 2007. *American Journal of Archaeology, 111,* 523-547.

Savage, S., Keller, D., Tuttle, C., 2008. *American Journal of Archaeology, 112,* 112-513.

Sayre, E.V., 1963. The intentional use of antimony and manganese in ancient glasses. In: Matson, F.R., Rindone, G. (eds.) *Advances in Glass Technology, Part 2.* Plenum Press, New York, 263-282.

Sayre, E.V., Smith, R.V., 1961. Compositional categories of ancient glass. *Science 133,* 1824-1826.

Schibille, N., 2011. Late Byzantine Mineral Soda High Alumina Glasses from Asia Minor: A New Primary Glass Production Group. *PLoS One 6,* 1-13.

Schibille, N., Marii, F., Rehren, T., 2008. *Archaeometry 50,* 627-642.

Schibille, N., Degryse, P., O'Hea, M., Izmer, A., Vanhaecke, F., Mc Kenzie, J., 2012. Late Roman glass form the 'Great Temple' at Petra and Khirbet et Tannur, Jordan – technology and provenance. *Archaeometry 54*, 997-1022.

Scrivner, A.E., Vance, D., Rohling, E.J., 2004. New neodymium isotope data quantify Nile involvement in Mediterranean anoxic episodes. *Geology 32*, 565-568.

Shahid, K.A., Glasser, F.P., 1972. Phase equilibria in the system Na-Ca-Mg-Si. *Physics and Chemistry of Glasses 13*, 27-42.

Shannon, R.D., 1976. Revised effective ionic radii and systematic studies of interatomic distances in halides and chalcogenides. *Acta Crystallographica A32*, 751-767.

Shelby, J.E., 2005. *Introduction to Glass Science and Technology, 2nd ed*. Royal Society of Chemistry, Cambridge.

Shortland, A.J., 2004. Evaporites of the Wadi Natrun: seasonal and annual variation and its implication for ancient exploitation. *Archaeometry 46*, 497-516.

Shortland, A.J., Schroeder H., 2009. Analysis of first millennium BC glass vessels and beads from the Pichvnari necropolis, Georgia. *Archaeometry 51*, 947-965.

Shortland, A.J., Schachner, L., Freestone, I.C., Tite, M., 2006. Natron as a flux in the early vitreous materials industry: sources, beginnings and reasons for decline. *Journal of Archaeological Science 33*, 521-530.

Shortland, A.J., Rogers, N., Eremin, K., 2007. Trace element discriminants between Egyptian and Mesopotamian late Bronze Age glasses. *Journal of Archaeological Science 34*, 781-789.

Shugar, A., Rehren, Th., 2002. Formation and composition of glass as a function of firing temperature. *Glass Technology 43C*, 145-150.

Silvestri, A., 2008. The coloured glass of Iulia Felix. *Journal of Archaeological Science 35*, 1489-1501.

Silvestri, A., Molin, G., Salviulo, G., 2005. Roman and Medieval glass from the Italian area: bulk characterization and relationships with production technologies. *Archaeometry 47*, 797-816.

Silvestri, A., Molin, G., Salviulo, G., Schievenin, R., 2006. Sand for Roman glass production: an experimental and philological study on source of supply. *Archaeometry 48*, 415-432.

Silvestri, A., Molin, G., Salviulo, G., 2008. The colourless glass of Iulia Felix. *Journal of Archaeological Science 35*, 331-341.

Sivan, D., Lambeck, K., Toueg, R., Raban, A., Porath, Y., Shirman, B., 2004. Ancient coastal wells of Caesarea Maritima, Israel, an indicator for relative sea level changes during the last 2000 years. *Earth and Planetary Science Letters 222*, 315-330.

Smedley, J.W., Jackson, C.M., 2002. Medieval and post-medieval glass technology: batch measuring practices. *Glass Technology 43*, 22-27.

Smedley, J.W., Jackson, C.M., Booth, C.A., 1998. Back to the roots: The raw materials, glass recipes and glassmaking practices of Theophilus. In: McCray, P., Kingery, W.D. (eds.) The prehistory and history of glass and glassmaking technology. *Ceramics and civilisation 8*, 145-165.

Sode, T., Kock, J., 2001. Traditional Raw Glass Production in Northern India: The Final Stage of an Ancient Technology. *J. Glass Stud. 43*, 155–169.

Stager, L.E., 1980. *The Rite of Child Sacrifice at Carthage, in New Light on Ancient Carthage*. J.G. Pedley (ed.). Ann Arbor, University of Michigan Press.

Stager, L.E., 1982. "Carthage: A View from the Tophet." in *Phonizier im Western*. H.G. Niemeyer (ed.), Madrider Beitrage 8, 155-166.

Stager, L.E., Wolff, S.R., 1984. Child Sacrifice at Carthage - Religious Rite or Population Control? Archaeological Evidence for a New Analysis, *Biblical Archaeology Review* 10:31-51.

Stager, L.E., Greene, J.A., Garnand, B.K., Birney, K. "Carthage: "The Precinct of Tanit (Tophet). The American Schools of Oriental Research Punic Project. The Stager Excavations (1976-1979)" in *Studies in the Archaeology and History of the Levant 6*, Semitic Museum Publications, forthcoming.

Stanley, D.J., Krom, M.D., Cliff, R.A., Woodward, J.C., 2003. Nile flow failure at the end of the Old Kingdom, Egypt: strontium isotopic and petrologic evidence. *Geoarchaeology 18*, 395-402.

Stern, E.M., 1999. Roman Glassblowing in a Cultural Context. *Am. J. Archaeol. 103*, 441–484.

Stern, E.M., 2002. Glass is Hot. *Am. J. Archaeol. 106*, 463–471.

Stevenson, C., 1914. *Bowl of Roman Madrepore Glass*. *Bull*. Pennsylvania Museum 12, 25–27.

Stille, P., Shields, G., 1997. Radiogenic Isotope Geochemistry of Sedimentary and Aquatic Systems. *Lecture Notes in Earth Sciences 68*, Springer, Berlin, 217.

Sun, S.-s., McDonough, W.F., 1989. Chemical and isotopic systematics of oceanic basalts: implications for mantle composition and processes. *Geological Society, London, Special Publications 42*, 313-345.

Tachikawa, K., Roy-Barman, M., Michard, A., Thouron, D., Yeghicheyan, D., Jeandel, C., 2004. Neodymium isotopes in the Mediterranean Sea: comparison between seawater and sediment signals. *Geochimica et Cosmochimica Acta 68*, 3095-3106.

Tal, O., Jackson-Tal, R.E., Freestone, I.C., 2004. New evidence of the production of raw glass at Late Byzantine Apollonia-Arsuf, Israel. *Journal of Glass Studies 46*, 51-66.

Tanimoto, S., Rehren, T., 2008. Interactions between silicate and salt melts in LBA glassmaking. *Journal of Archaeological Science 35*, 2566-2573.

Thorley, J., 1969. *The Development of Trade between the Roman Empire and the East under Augustus*. Greece Rome 16, 209–223.

Tiede, R.L., Tooley, F.V., 1945. Effect of temperature on homogenising rate of Na_2O CaO SiO_2 glass. *Journal of the American Ceramic Society 28*, 42-47.

Tlig, S., Steinberg, M., 1982. Distribution of rare-earth elements (REE) in size fractions of recent sediments of the Indian Ocean. *Chemical Geology 37*, 317-333.

Toniolo A., 2003. Il carico di rottami di vetro del relitto di Grado. In: Il vetronell'Alto Adriatico. *Atti delle IX Giornate Nazionali di Studio Comitato Nazionale AIIV*.

Turner, W.E.S., 1956. Studies of ancient glass and glassmaking processes. Part V: Raw materials and melting processes. *Journal of the Society of Glass Technology 40*, 276-299.

Vallotto, M., Verità, M., 2002. Glasses from Pompeii and Herculaneum and the sands of the Rivers Belus and Volturno. In: Renn, J., Castagnetti, G. (eds.) *Homo Faber: Studies on Nature, Technology and Science at the Time of Pompeii*. Presented at a Conference at the Deutsches Museum Munich, 21-22 March 2000. L'Erma di Bretschneider, Rome, 63-73.

Vanhoutte, S., 2008. *Relicta, Archaeologie, Monumenten en Landschapsonderzoek in Vlaanderen, 3*, 199-236.

Vanhoutte, S., Bastiaens, J., De Clercq, W., Deforce, K., Ervynck, A., Fret, M., Haneca, K., Lentacker, A., Stieperaere, H., Van Neer, W., Cosyns, P., Degryse, P., Dhaeze, W., Dijkman, W., Lyne, M., Rogers, P., Van Driel-Murray, C., Van Heesch, J., Wild, J. P., 2009. *Relicta, Archaeologie, Monumenten en Landschapsonderzoek in Vlaanderen, 5*, 9-140.

Veizer, J., 1977. Diagenesis of pre-Quaternary carbonates as indicated by tracer studies. *Journal of Sedimentary Research 47*, 565-581.

Vella, C., Fleury, T.-J., Raccasi, G., Provansal, M., Sabatier, F., Bourcier, M., 2005. Evolution of the Rhône delta plain in the Holocene. Marine Geology 222-223, 235-265.

Vermeulen, F., Verhoeven, G., 2006. An integrated survey of Roman urbanization at Potentia, Central Italy. *Journal of Field Archaeology 31*, 395-410.

Vermeulen, F., Hay, S., Verhoeven, G., Baldwin, E., Ferraby, R., Verdonck, L., de Seranno, S., 2006. Potentia: an integrated survey of a Roman colony on the Adriatic coast. *Papers of the British School at Rome 74*, 203-236.

Viñals, M.J., Fumanales, M.P., 1995. Quaternary development and evolution of the sedimentary environments in the central Mediterranean Spanish coast. *Quaternary International* 29-30, 119-128.

Vlachou, C., McDonnell, J.G., Janaway, R.C., 2002. Experimental Investigation of Silvering in Late Roman Coinage. *Mater. Res. Soc. Symp. Proc. 712*, 119.2.1–119.2.9.

Walton, M.S., Shortland, A., Kirk, S., Degryse, P., 2009. Evidence for the trade of Mesopotamian and Egyptian glass to Mycenaean Greece. *Journal of Archaeological Science 36*, 1496-1503.

Wang, X., Griffin, W.L., Chen, J., Huang, P., Li, X., 2011. U and Th contents and Th/U ratios of zircon in felsic and mafic magmatic rocks: improved zircon-melt distribution coefficients. *Acta Geologica Sinica 85*, 164-174.

Wedepohl, K.H., 1978. *Handbook of Geochemistry Vol. II/4 Section 38, Strontium.* Springer-Verlag, Berlin.

Wedepohl, K.H., 1995. The composition of the continental crust. *Geochimica et Cosmochimica Acta 59*, 1217-1232.

Wedepohl, K.H., Baumann, A., 2000. The use of marine molluskan shells for Roman glass and local raw glass production in the Eifel area (Western Germany). *Naturwissenschaften 87*, 129-132.

Wedepohl, K.H., Simon, K., Kronz, A., 2011a. Data on 61 chemical elements for the characterization of three major glass compositions in Late Antiquity and the Middle Ages. *Archaeometry 53*, 81-102.

Wedepohl, K.H., Simon, K., Kronz, A., 2011b. The chemical composition including the Rare Earth Elements of the three major glass types of Europe and the Orient used in late antiquity and the Middle Ages. *Chemie der Erde 71*, 289-296.

Weldeab, S., Emeis, K.C., Hemleben, C., Siebel, W., 2002. Provenance of lithogenic surface sediments and pathways of riverine suspended matter in the eastern Mediterranean Sea: evidence from $^{143}Nd/^{144}Nd$ and $^{87}Sr/^{86}Sr$ ratios. *Chemical Geology 186*, 139-149.

West, L.C., 1932. The Economic Collapse of the Roman Empire. *Class. J. 28*, 96–106.

Weyl, W.A., 1951. *Coloured glasses.* Society of Glass Technology, Sheffield.

Wilson, L., Pollard, A.M., 2001. The provenance hypothesis. In: Brothwell, D.R., Pollard, A.M. (eds.) *Handbook of Archaeological Sciences*. John Wiley and sons, Chichester, 507-517.

Yokoo, Y., Nakano, T., Nishikawa, M., Quan, H., 2004. Mineralogical variation of Sr–Nd isotopic and elemental compositions in loess and desert sand from the central Loess Plateau in China as a provenance tracer of wet and dry deposition in the northwestern Pacific. *Chemical Geology 204*, 45-62.

Loeb classical library refs

Athenaeus. The Learned Banqueters. Volume 5: Books 10.420e-11. Translated by Olson, S.D., 2009. Loeb Classical Library 274. Cambridge, MA: Harvard University Press

Historia Augusta. Volume 2: 18 Severus Alexander (24). Translated by Magie, D., 1924. Loeb Classical Library 140. Cambridge, MA: Harvard University Press.

Historia Augusta. Volume 3: 26 The Deified Aurelian (45). Translated by Magie, D., 1932. Loeb Classical Library 263. Cambridge, MA: Harvard University Press.

Historia Augusta. Volume 3: 29 Firmus, Saturninus, Proculus and Bonosus (9). Translated by Magie, D., 1932. Loeb Classical Library 263. Cambridge, MA: Harvard University Press.

Josephus. The Jewish War. Volume 1: Books 1-2. Translated by Thackery, H. St. J., 1927. Loeb Classical Library 203. Cambridge, MA: Harvard University Press.

Pliny. Natural History. Volume 8: Books 28-32. Translated by Jones, W.H.S., 1963. Loeb Classical Library 418. Cambridge, MA: Harvard University Press.

Pliny. Natural History. Volume 10: Books 36-37. Translated by Eichholz, D.E., 1962. Loeb Classical Library 419. Cambridge, MA: Harvard University Press.

Strabo. Geography. Volume 5: Books 10-12. Translated by Jones, H.L., 1928. Loeb Classical Library 211. Cambridge, MA: Harvard University Press.

Strabo. Geography. Volume 7: Books 15-16. Translated by Jones, H.L., 1930. Loeb Classical Library 241. Cambridge, MA: Harvard University Press.

Strabo. Geography. Volume 8: Book 17. Translated by Jones, H.L., 1932. Loeb Classical Library 267. Cambridge, MA: Harvard University Press.

Tacitus. Histories: Books 4-5. Translated by Moore, C.H., Jackson, J., 1931. Loeb Classical Library 249. Cambridge, MA: Harvard University Press.

Tacitus. Histories. Translated by Church, A.J., Brodribb, W.J., 1864. Cambridge: Macmillan and Co.

9 789462 700079